Standards Practice Book
For Home or School
Grade 3

Houghton
Mifflin
Harcourt

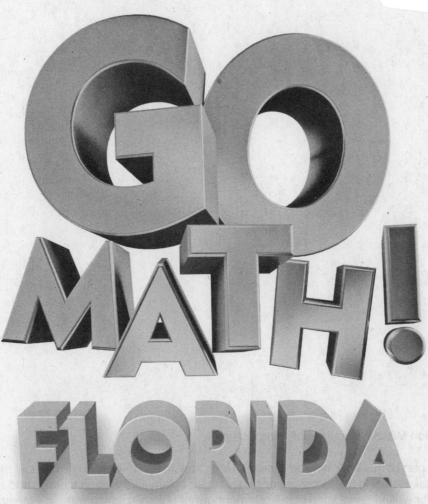

INCLUDES:

- Home or School Practice
- Lesson Practice and Test Preparation
- English and Spanish School-Home Letters
- Getting Ready for Grade 4 Lessons

Copyright © by Houghton Mifflin Harcourt Publishing Company

All rights reserved. No part of this work may be reproduced or transmitted in any form or by any means, electronic or mechanical, including photocopying or recording, or by any information storage and retrieval system, without the prior written permission of the copyright owner unless such copying is expressly permitted by federal copyright law. Requests for permission to make copies of any part of the work should be addressed to Houghton Mifflin Harcourt Publishing Company, Attn: Contracts, Copyrights, and Licensing, 9400 Southpark Center Loop, Orlando, Florida 32819-8647.

Printed in the U.S.A.

ISBN 978-0-544-50170-6

4 5 6 7 8 9 10   0982   23 22 21 20 19 18 17

4500652023   ^ B C D E F G

If you have received these materials as examination copies free of charge, Houghton Mifflin Harcourt Publishing Company retains title to the materials and they may not be resold. Resale of examination copies is strictly prohibited.

Possession of this publication in print format does not entitle users to convert this publication, or any portion of it, into electronic format.

Whole Number Operations

Developing understanding of multiplication and division and strategies for multiplication and division within 100

1 Addition and Subtraction within 1,000

Domains Operations and Algebraic Thinking
Number and Operations in Base Ten

2 Represent and Interpret Data

Domain Measurement and Data

© Houghton Mifflin Harcourt Publishing Company

3 Understand Multiplication

Domain Operations and Algebraic Thinking

4 Multiplication Facts and Strategies

Domain Operations and Algebraic Thinking

© Houghton Mifflin Harcourt Publishing Company

5 Use Multiplication Facts

Domains Operations and Algebraic Thinking
Number and Operations in Base Ten

6 Understand Division

Domain Operations and Algebraic Thinking

Division Facts and Strategies

Domain Operations and Algebraic Thinking

© Houghton Mifflin Harcourt Publishing Company

Fractions

Developing understanding of fractions, especially unit fractions
(fractions with numerator 1)

© Houghton Mifflin Harcourt Publishing Company

Measurement

Developing understanding of the structure of rectangular arrays and of area

10 Time, Length, Liquid Volume, and Mass

Domain Measurement and Data

11 Perimeter and Area

Domain Measurement and Data

© Houghton Mifflin Harcourt Publishing Company

Geometry

Describing and analyzing two-dimensional shapes

12   Two-Dimensional Shapes

Domain Geometry

© Houghton Mifflin Harcourt Publishing Company

End-of-Year Resources

Getting Ready for Grade 4

These lessons review important skills and prepare you for Grade 4.

© Houghton Mifflin Harcourt Publishing Company

Chapter 1 School-Home Letter

Vocabulary

estimate A number close to an exact amount

compatible numbers Numbers that are easy to compute mentally and are close to the real numbers

Dear Family,

During the next few weeks, our math class will be learning to estimate and solve addition and subtraction problems using numbers through hundreds.

You can expect to see homework that provides practice with adding and subtracting numbers as well as estimating sums and differences.

Here is a sample of how your child will be taught to estimate sums.

🔑 MODEL Estimate Sums

These are two methods we will be using to estimate sums.

$367 + 512 = \blacksquare$

Use rounding.		**Use compatible numbers.**	
STEP 1	**STEP 2**	**STEP 1**	**STEP 2**
Round each number to the nearest hundred.	Add the rounded numbers.	Find a compatible number for each addend.	Add the numbers mentally.
$367 \rightarrow 400$ $+\,512 \rightarrow 500$	400 $+\,500$ 900	$105 \rightarrow 100$ $+\,362 \rightarrow 400$	100 $+\,400$ 500

Tips

Choosing Compatible Numbers to Estimate Sums and Differences

A number may have more than one compatible number. For example, a compatible number for 362 could be 350 or 400. Whichever numbers are easiest to add or subtract mentally are the best ones to use for estimations.

Activity

Provide books with large numbers of pages (3-digit numbers). Have your child use rounding and compatible numbers to estimate the total number of pages in the two books and compare how many more pages one book has than the other.

© Houghton Mifflin Harcourt Publishing Company

Carta
para la casa

Vocabulario

estimación Un número que se aproxima a una cantidad exacta

números compatibles Números con los que es fácil calcular mentalmente y que se aproximan a los números reales

Querida familia,

Durante las próximas semanas, en la clase de matemáticas aprenderemos a estimar y resolver problemas de suma y resta usando números hasta las centenas.

Llévare a la casa tareas con actividades para practicar la suma y la resta, y para estimar sumas y diferencias.

Este es un ejemplo de la manera como aprenderemos a estimar sumas.

🔒 MODELO Estimar sumas

Estos son dos métodos que usaremos para estimar sumas.

$367 + 512 = \blacksquare$

Usa el redondeo.

PASO 1	PASO 2
Redondea cada número a la centena más cercana.	Suma los números que hallaste.

$$367 \rightarrow 400$$
$$+\ 512 \rightarrow 500$$

$$400$$
$$+\ 500$$
$$\overline{900}$$

Usa números compatibles.

PASO 1	PASO 2
Halla un número compatible para cada sumando.	Suma los números mental-mente.

$$105 \rightarrow 100$$
$$+\ 362 \rightarrow 400$$

$$100$$
$$+\ 400$$
$$\overline{500}$$

Pistas

Elegir números compatibles para estimar sumas y restas

Un número puede tener más de un número compatible. Por ejemplo, un número compatible para 362 puede ser 350 o 400. Cualquiera de los números con el que sea más fácil sumar y restar mentalmente sirve para hacer estimaciones.

Actividad

Dé a su hijo o hija dos libros que tengan bastantes páginas (con números de 3 dígitos). Pídale que use el redondeo y los números compatibles para estimar el total de páginas de los dos libros y para averiguar cuántas más páginas tiene un libro que el otro.

© Houghton Mifflin Harcourt Publishing Company

Name _____

Number Patterns

Find the sum. Then use the Commutative Property of Addition to write the related addition sentence.

1. 9 + 2 = **11**   **4.** 3 + 10 = ____   **7.** 8 + 9 = ____

2 + **9** = **11**   ____ + ____ = ____   ____ + ____ = ____

2. 4 + 7 = ____   **5.** 6 + 7 = ____   **8.** 0 + 4 = ____

____ + ____ = ____   ____ + ____ = ____   ____ + ____ = ____

3. 3 + 6 = ____   **6.** 7 + 5 = ____   **9.** 9 + 6 = ____

____ + ____ = ____   ____ + ____ = ____   ____ + ____ = ____

Is the sum even or odd? Write *even* or *odd*.

10. 5 + 2 _____   **11.** 6 + 4 _____   **12.** 1 + 0 _____

13. 5 + 5 _____   **14.** 3 + 8 _____   **15.** 7 + 7 _____

Problem Solving REAL WORLD

16. Ada writes 10 + 8 = 18 on the board. Maria wants to use the Commutative Property of Addition to rewrite Ada's addition sentence. What number sentence should Maria write?

_____

17. Jackson says he has an odd number of model cars. He has 6 cars on one shelf and 8 cars on another shelf. Is Jackson correct? **Explain.**

_____

_____

_____

© Houghton Mifflin Harcourt Publishing Company

Lesson Check

1. Marvella says that the sum of her addends is odd. Which of the following could be Marvella's addition problem?

(A) 5 + 3   (C) 2 + 8

(B) 9 + 7   (D) 5 + 6

2. Which number sentence shows the Commutative Property of Addition?

$$3 + 9 = 12$$

(A) 12 − 9 = 3   (C) 9 + 3 = 12

(B) 12 = 8 + 4   (D) 12 − 3 = 9

Spiral Review

3. Amber has 2 quarters, a dime, and 3 pennies. How much money does Amber have? (Grade 2)

(A) 53¢   (C) 63¢

(B) 58¢   (D) 68¢

4. Josh estimates the height of his desk. Which is the best estimate? (Grade 2)

(A) 1 foot   (C) 5 feet

(B) 2 feet   (D) 9 feet

Use the bar graph for 5–6.

5. Who read the most books? (Grade 2)

(A) Alicia

(B) Bob

(C) Juan

(D) Maria

6. Who read 3 more books than Bob? (Grade 2)

(A) Alicia

(B) Juan

(C) Maria

(D) no one

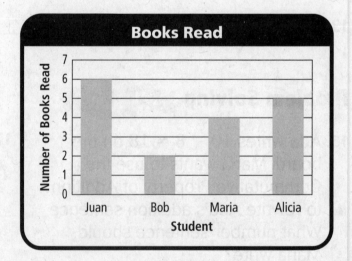

© Houghton Mifflin Harcourt Publishing Company

Name _____

Round to the Nearest Ten or Hundred

Locate and label 739 on the number line. Round to the nearest hundred.

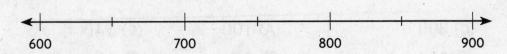

600   700   800   900

1. 739 is between ___700___ and ___800___.

2. 739 is closer to _____ than it is to _____.

3. 739 rounded to the nearest hundred is _____.

Round to the nearest ten and hundred.

4. 363 _____   **5.** 829 _____   **6.** 572 _____

_____   _____   _____

7. 209 _____   **8.** 663 _____   **9.** 949 _____

_____   _____   _____

10. 762 _____   **11.** 399 _____   **12.** 402 _____

_____   _____   _____

Problem Solving REAL WORLD

13. The baby elephant weighs 435 pounds. What is its weight rounded to the nearest hundred pounds?

_____

14. Jayce sold 218 cups of lemonade at his lemonade stand. What is 218 rounded to the nearest ten?

_____

© Houghton Mifflin Harcourt Publishing Company

Lesson Check

1. One day, 758 people visited the Monkey House at the zoo. What is 758 rounded to the nearest hundred?

- Ⓐ 700
- Ⓒ 800
- Ⓑ 760
- Ⓓ 860

2. Sami ordered 132 dresses for her store. What is 132 rounded to the nearest ten?

- Ⓐ 100
- Ⓒ 140
- Ⓑ 130
- Ⓓ 200

Spiral Review

3. Which describes the number sentence? (Lesson 1.1)

$$6 + 0 = 6$$

- Ⓐ Commutative Property of Addition
- Ⓑ Identity Property of Addition
- Ⓒ even + odd = odd
- Ⓓ odd + odd = odd

4. Which has an even sum? (Lesson 1.1)

- Ⓐ 7 + 4
- Ⓑ 2 + 6
- Ⓒ 5 + 4
- Ⓓ 3 + 2

5. What name describes this shape? (Grade 2)

- Ⓐ cone
- Ⓑ cube
- Ⓒ rectangle
- Ⓓ triangle

6. What word describes the equal shares of the shape? (Grade 2)

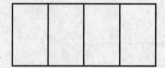

- Ⓐ wholes
- Ⓑ thirds
- Ⓒ halves
- Ⓓ fourths

© Houghton Mifflin Harcourt Publishing Company

Name _____

Estimate Sums

Use rounding or compatible numbers to estimate the sum.

1. $\begin{array}{r} 198 \\ + 727 \\ \hline \end{array}$ $\begin{array}{r} \mathbf{200} \\ + \mathbf{725} \\ \hline \mathbf{925} \end{array}$

2. $\begin{array}{r} 87 \\ + 34 \\ \hline \end{array}$ $\begin{array}{r} \\ + \\ \hline \end{array}$

3. $\begin{array}{r} 222 \\ + 203 \\ \hline \end{array}$ $\begin{array}{r} \\ + \\ \hline \end{array}$

4. $\begin{array}{r} 52 \\ + 39 \\ \hline \end{array}$ $\begin{array}{r} \\ + \\ \hline \end{array}$

5. $\begin{array}{r} 256 \\ + 321 \\ \hline \end{array}$ $\begin{array}{r} \\ + \\ \hline \end{array}$

6. $\begin{array}{r} 302 \\ + 412 \\ \hline \end{array}$ $\begin{array}{r} \\ + \\ \hline \end{array}$

7. $\begin{array}{r} 519 \\ + 124 \\ \hline \end{array}$ $\begin{array}{r} \\ + \\ \hline \end{array}$

8. $\begin{array}{r} 790 \\ + 112 \\ \hline \end{array}$ $\begin{array}{r} \\ + \\ \hline \end{array}$

9. $\begin{array}{r} 547 \\ + 326 \\ \hline \end{array}$ $\begin{array}{r} \\ + \\ \hline \end{array}$

10. $325 + 458$

_____ + _____ = _____

11. $620 + 107$

_____ + _____ = _____

Problem Solving REAL WORLD

12. Stephanie read 72 pages on Sunday and 83 pages on Monday. About how many pages did Stephanie read during the two days?

_____

13. Matt biked 345 miles last month. This month he has biked 107 miles. Altogether, about how many miles has Matt biked last month and this month?

_____

© Houghton Mifflin Harcourt Publishing Company

Lesson Check

1. The McBrides drove 317 miles on one day and 289 on the next day. What is the best estimate of the number of miles the McBrides drove in all during the two days?

Ⓐ 100
Ⓑ 400
Ⓒ 500
Ⓓ 600

2. Ryan counted 63 birds in his backyard last week. This week, he counted 71 birds in his backyard. About how many birds did Ryan count in all?

Ⓐ about 70
Ⓑ about 100
Ⓒ about 130
Ⓓ about 200

Spiral Review

3. What name describes this shape?

(Grade 2)

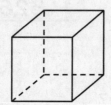

Ⓐ cone
Ⓑ cube
Ⓒ quadrilateral
Ⓓ square

4. Which has an odd sum? (Lesson 1.1)

Ⓐ 9 + 9
Ⓑ 5 + 3
Ⓒ 6 + 7
Ⓓ 2 + 8

5. What is 503 rounded to the nearest hundred? (Lesson 1.2)

Ⓐ 500
Ⓑ 510
Ⓒ 600
Ⓓ 610

6. What is 645 rounded to the nearest ten? (Lesson 1.2)

Ⓐ 600
Ⓑ 640
Ⓒ 650
Ⓓ 700

© Houghton Mifflin Harcourt Publishing Company

Mental Math Strategies for Addition

Count by tens and ones to find the sum.
Use the number line to show your thinking.

1. $29 + 14 =$ __**43**__

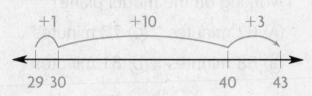

2. $36 + 28 =$ _____

3. $45 + 26 =$ _____

4. $52 + 34 =$ _____

Use mental math to find the sum.
Draw or describe the strategy you use.

5. $52 + 19 =$ _____

6. $122 + 306 =$ _____

Problem Solving REAL WORLD

7. Shelley spent 17 minutes washing the dishes. She spent 38 minutes cleaning her room. **Explain** how you can use mental math to find how long Shelley spent on the two tasks.

_____

_____

_____

8. It took Marty 42 minutes to write a book report. Then he spent 18 minutes correcting his report. **Explain** how you can use mental math to find how long Marty spent on his book report.

_____

_____

_____

© Houghton Mifflin Harcourt Publishing Company

Lesson Check

1. Sylvia spent 36¢ for a pencil and 55¢ for a notepad. Use mental math to find how much she spent in all.

Ⓐ 80¢   Ⓒ 90¢

Ⓑ 81¢   Ⓓ 91¢

2. Will spent 24 minutes putting together a model plane. Then he spent 48 minutes painting the model. How long did Will spend working on the model plane?

Ⓐ 62 minutes   Ⓒ 72 minutes

Ⓑ 68 minutes   Ⓓ 81 minutes

Spiral Review

3. What name describes this shape?

(Grade 2)

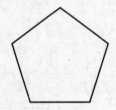

Ⓐ hexagon

Ⓑ pentagon

Ⓒ quadrilateral

Ⓓ triangle

4. What word describes the equal shares of the shape? (Grade 2)

Ⓐ fourths

Ⓑ halves

Ⓒ sixths

Ⓓ thirds

5. Tammy wrote an addition problem that has an odd sum. Which could be Tammy's addition problem?

(Lesson 1.1)

Ⓐ 2 + 6

Ⓑ 3 + 5

Ⓒ 5 + 6

Ⓓ 7 + 7

6. Greg counted 83 cars and 38 trucks in the mall parking lot. Which is the best estimate of the total number of cars and trucks Greg counted? (Lesson 1.3)

Ⓐ 100

Ⓑ 110

Ⓒ 120

Ⓓ 130

© Houghton Mifflin Harcourt Publishing Company

Name _____

Use Properties to Add

Use addition properties and strategies to find the sum.

1. 34 + 62 + 51 + 46 = __193__

$$\begin{array}{r} 34 \\ 46 \\ 62 \\ +\ 51 \\ \hline 193 \end{array}$$

10 ⟩ 10
10 ⟨

2. 27 + 68 + 43 = _____

3. 42 + 36 + 18 = _____

4. 74 + 35 + 16 + 45 = _____

5. 41 + 26 + 149 = _____

6. 52 + 64 + 28 + 44 = _____

Problem Solving REAL WORLD

7. A pet shelter has 26 dogs, 37 cats, and 14 gerbils. How many of these animals are in the pet shelter in all?

8. The pet shelter bought 85 pounds of dog food, 50 pounds of cat food, and 15 pounds of gerbil food. How many pounds of animal food did the pet shelter buy?

© Houghton Mifflin Harcourt Publishing Company

Lesson Check

1. At summer camp there are 52 boys, 47 girls, and 18 adults. How many people are at summer camp?

(A) 97

(B) 107

(C) 117

(D) 127

2. At camp, 32 children are swimming, 25 are fishing, and 28 are canoeing. How many children are swimming, fishing, or canoeing?

(A) 75

(B) 85

(C) 95

(D) 105

Spiral Review

3. Four students estimated the width of the door to their classroom. Who made the best estimate? (Grade 2)

(A) Ted: 1 foot

(B) Hank: 3 feet

(C) Ann: 10 feet

(D) Maria: 15 feet

4. Four students estimated the height of the door to their classroom. Who made the best estimate? (Grade 2)

(A) Larry: 1 meter

(B) Garth: 2 meters

(C) Ida: 14 meters

(D) Jill: 20 meters

5. Jeff's dog weighs 76 pounds. What is the dog's weight rounded to the nearest ten pounds? (Lesson 1.2)

(A) 70 pounds

(B) 80 pounds

(C) 90 pounds

(D) 100 pounds

6. Ms. Kirk drove 164 miles in the morning and 219 miles in the afternoon. Which is the best estimate of the total number of miles she drove that day? (Lesson 1.3)

(A) 100 miles

(B) 200 miles

(C) 400 miles

(D) 500 miles

© Houghton Mifflin Harcourt Publishing Company

Use the Break Apart Strategy to Add

Estimate. Then use the break apart strategy to find the sum.

1. Estimate: __800__

$$\begin{array}{r} 325 \\ + \ 494 \\ \hline \end{array} \begin{array}{l} = \ 300 + 20 + 5 \\ = \ 400 + 90 + 4 \\ \hline \ \ \ 700 + 110 + 9 \end{array}$$

2. Estimate: _____

$$\begin{array}{r} 518 \ = \\ + \ 372 \ = \\ \hline \end{array}$$

3. Estimate: _____

$$\begin{array}{r} 473 \ = \\ + \ 123 \ = \\ \hline \end{array}$$

4. Estimate: _____

$$\begin{array}{r} 208 \ = \\ + \ 569 \ = \\ \hline \end{array}$$

5. Estimate: _____

$$\begin{array}{r} 731 \ = \\ + \ 207 \ = \\ \hline \end{array}$$

6. Estimate: _____

$$\begin{array}{r} 495 \ = \\ + \ 254 \ = \\ \hline \end{array}$$

Problem Solving REAL WORLD

Use the table for 7–8.

7. Laura is making a building using Set A and Set C. How many blocks can she use in her building?

_____

8. Clark is making a building using Set B and Set C. How many blocks can he use in his building?

_____

Build-It Blocks	
Set	Number of Blocks
A	165
B	188
C	245

© Houghton Mifflin Harcourt Publishing Company

Lesson Check

1. Arthur read two books last week. One book has 216 pages. The other book has 327 pages. Altogether, how many pages are in the two books?

   (A) 533
   (B) 543
   (C) 633
   (D) 643

2. One skeleton in a museum has 189 bones. Another skeleton has 232 bones. How many bones in all are in the two skeletons?

   (A) 311
   (B) 312
   (C) 411
   (D) 421

Spiral Review

3. Culver has 1 quarter, 3 dimes, and a penny. How much money does he have? (Grade 2)

   (A) 41¢
   (B) 55¢
   (C) 56¢
   (D) 86¢

4. Felicia has 34 quarters, 25 dimes, and 36 pennies. How many coins does Felicia have? (Lesson 1.5)

   (A) 75
   (B) 85
   (C) 95
   (D) 105

5. Jonas wrote $9 + 8 = 17$. Which number sentence shows the Commutative Property of Addition?

   (Lesson 1.1)

   (A) $9 + 0 = 9$
   (B) $8 + 9 = 17$
   (C) $17 - 9 = 8$
   (D) $17 - 8 = 9$

6. At Kennedy School there are 37 girls and 36 boys in the third grade. How many students are in the third grade at Kennedy School?

   (Lesson 1.4)

   (A) 63
   (B) 73
   (C) 81
   (D) 91

© Houghton Mifflin Harcourt Publishing Company

Use Place Value to Add

Estimate. Then find the sum.

1. Estimate: **600**

$$\begin{array}{r} \overset{1}{3}24 \\ +\ 285 \\ \hline \textbf{609} \end{array}$$

2. Estimate: _____

$$\begin{array}{r} 519 \\ +\ 347 \\ \hline \end{array}$$

3. Estimate: _____

$$\begin{array}{r} 323 \\ +\ 151 \\ \hline \end{array}$$

4. Estimate: _____

$$\begin{array}{r} 169 \\ +\ 354 \\ \hline \end{array}$$

5. Estimate: _____

$$\begin{array}{r} 148 \\ +\ 285 \\ \hline \end{array}$$

6. Estimate: _____

$$\begin{array}{r} 270 \\ +\ 453 \\ \hline \end{array}$$

7. Estimate: _____

$$\begin{array}{r} 275 \\ +\ 116 \\ \hline \end{array}$$

8. Estimate: _____

$$\begin{array}{r} 157 \\ +\ 141 \\ \hline \end{array}$$

9. Estimate: _____

$$\begin{array}{r} 127 \\ +\ 290 \\ \hline \end{array}$$

10. Estimate: _____

$$\begin{array}{r} 258 \\ +\ 565 \\ \hline \end{array}$$

11. Estimate: _____

$$\begin{array}{r} 311 \\ +\ 298 \\ \hline \end{array}$$

12. Estimate: _____

$$\begin{array}{r} 534 \\ +\ 256 \\ \hline \end{array}$$

Problem Solving REAL WORLD

13. Mark has 215 baseball cards. Emily has 454 baseball cards. How many baseball cards do Mark and Emily have altogether?

_____

14. Jason has 330 pennies. Richie has 268 pennies. Rachel has 381 pennies. Which two students have more than 700 pennies combined?

_____

© Houghton Mifflin Harcourt Publishing Company

Lesson Check

1. There are 167 students in the third grade. The same number of students is in the fourth grade. How many third graders and fourth graders are there in all?

   Ⓐ 224
   Ⓑ 234
   Ⓒ 324
   Ⓓ 334

2. Jamal read a book with 128 pages. Then he read a book with 179 pages. How many pages did Jamal read in all?

   Ⓐ 397
   Ⓑ 307
   Ⓒ 297
   Ⓓ 207

Spiral Review

3. Adam travels 248 miles on Monday. He travels 167 miles on Tuesday. Which is the best estimate for the total number of miles Adam travels? (Lesson 1.3)

   Ⓐ 200
   Ⓑ 300
   Ⓒ 400
   Ⓓ 500

4. Wes made $14, $62, $40, and $36 mowing lawns. How much did he make in all mowing lawns? (Lesson 1.5)

   Ⓐ $116
   Ⓑ $152
   Ⓒ $166
   Ⓓ $188

5. There are 24 students in Mrs. Cole's class and 19 students in Mr. Garmen's class. How many students in all are in the two classes? (Lesson 1.4)

   Ⓐ 43
   Ⓑ 40
   Ⓒ 33
   Ⓓ 5

6. There were 475 children at the baseball game on Sunday. What is 475 rounded to the nearest ten? (Lesson 1.2)

   Ⓐ 400
   Ⓑ 470
   Ⓒ 480
   Ⓓ 500

© Houghton Mifflin Harcourt Publishing Company

Name _____

Estimate Differences

Use rounding or compatible numbers to estimate the difference.

1. 40   **40**   2. 762 _____   3. 823 _____
  − 13   − **10**   − 332 − _____   − 242 − _____
           30

4. 98 _____   5. 287 _____   6. 359 _____
  − 49 − _____   − 162 − _____   − 224 − _____

7. 68 _____   8. 476 _____   9. 622 _____
  − 31 − _____   − 155 − _____   − 307 − _____

10. 771 − 531   11. 299 − 61

  _____ − _____ = _____   _____ − _____ = _____

Problem Solving

12. Ben has a collection of 812 stamps. He gives his brother 345 stamps. About how many stamps does Ben have left?

13. Savannah's bakery sold 284 pies in September. In October the bakery sold 89 pies. About how many more pies did Savannah's bakery sell in September than in October?

_____   _____

© Houghton Mifflin Harcourt Publishing Company

Lesson Check

1. Jorge has 708 baseball cards and 394 basketball cards. About how many more baseball cards than basketball cards does Jorge have?

   Ⓐ about 200

   Ⓑ about 300

   Ⓒ about 400

   Ⓓ about 500

2. Danika is making necklaces. She has 512 silver beads and 278 blue beads. About how many more silver than blue beads does Danika have?

   Ⓐ about 200

   Ⓑ about 300

   Ⓒ about 400

   Ⓓ about 800

Spiral Review

3. A store manager ordered 402 baseball caps and 122 ski caps. Which is the best estimate of the total number of caps the manager ordered? (Lesson 1.3)

   Ⓐ 300

   Ⓑ 500

   Ⓒ 600

   Ⓓ 700

4. Autumn collected 129 seashells at the beach. What is 129 rounded to the nearest ten? (Lesson 1.2)

   Ⓐ 100

   Ⓑ 120

   Ⓒ 130

   Ⓓ 200

5. Find the sum. (Lesson 1.7)

$$585 + 346$$

   Ⓐ 239

   Ⓑ 821

   Ⓒ 900

   Ⓓ 931

6. Julie made $22, $55, $38, and $25 babysitting. How much did she make in all babysitting? (Lesson 1.5)

   Ⓐ $102

   Ⓑ $115

   Ⓒ $140

   Ⓓ $165

© Houghton Mifflin Harcourt Publishing Company

Name _____

Mental Math Strategies
for Subtraction

Use mental math to find the difference.
Draw or describe the strategy you use.

1. $74 - 39 =$ ___**35**___

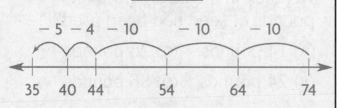

$$-5 \quad -4 \quad -10 \qquad -10 \qquad -10$$

35 40 44   54   64   74

2. $93 - 28 =$ _____

3. $51 - 9 =$ _____

4. $76 - 23 =$ _____

5. $357 - 214 =$ _____

6. $285 - 99 =$ _____

Problem Solving

7. Ruby has 78 books. Thirty-one of the books are on shelves. The rest are still packed in boxes. How many of Ruby's books are still in boxes?

8. Kyle has 130 pins in his collection. He has 76 of the pins displayed on his wall. The rest are in a drawer. How many of Kyle's pins are in a drawer?

_____

_____

© Houghton Mifflin Harcourt Publishing Company

Lesson Check

1. One day, a baker made 54 fruit pies. At the end of the day, only 9 of the pies were NOT sold. How many pies were sold that day?

   (A) 43   (C) 63

   (B) 45   (D) 65

2. George's father bought a 50-pound bag of wild bird seed. At the end of two weeks, 36 pounds of seed were left in the bag. How many pounds of seed had been used?

   (A) 14 pounds   (C) 26 pounds

   (B) 24 pounds   (D) 86 pounds

Spiral Review

3. For a party, Shaun blew up 36 red balloons, 28 white balloons, and 24 blue balloons. How many balloons did he blow up in all? (Lesson 1.5)

   (A) 78

   (B) 81

   (C) 87

   (D) 88

4. Tiffany has read 115 pages of her book. She has 152 pages left to read. How many pages are in the book? (Lesson 1.6)

   (A) 37

   (B) 267

   (C) 277

   (D) 367

5. The flower shop had 568 flowers on Monday. By Tuesday, the shop had 159 flowers left. About how many flowers had been sold? (Lesson 1.8)

   (A) about 200

   (B) about 300

   (C) about 400

   (D) about 500

6. There are 383 books in one section of the school library. Of the books, 165 are fiction books. Which is the best estimate of the number of books in that section that are NOT fiction? (Lesson 1.8)

   (A) about 200

   (B) about 300

   (C) about 400

   (D) about 500

© Houghton Mifflin Harcourt Publishing Company

Name _____

Use Place Value to Subtract

Estimate. Then find the difference.

1. Estimate: **500**

$$\begin{array}{r} {}^{7\ 15} \\ 5\cancel{8}5 \\ -\ 119 \\ \hline \end{array}$$

2. Estimate: _____

$$\begin{array}{r} 738 \\ -\ 227 \\ \hline \end{array}$$

3. Estimate: _____

$$\begin{array}{r} 651 \\ -\ 376 \\ \hline \end{array}$$

4. Estimate: _____

$$\begin{array}{r} 815 \\ -\ 281 \\ \hline \end{array}$$

5. Estimate: _____

$$\begin{array}{r} 487 \\ -\ 290 \\ \hline \end{array}$$

6. Estimate: _____

$$\begin{array}{r} 936 \\ -\ 329 \\ \hline \end{array}$$

7. Estimate: _____

$$\begin{array}{r} 270 \\ -\ 128 \\ \hline \end{array}$$

8. Estimate: _____

$$\begin{array}{r} 364 \\ -\ 177 \\ \hline \end{array}$$

9. Estimate: _____

$$\begin{array}{r} 627 \\ -\ 253 \\ \hline \end{array}$$

10. Estimate: _____

$$\begin{array}{r} 862 \\ -\ 419 \\ \hline \end{array}$$

11. Estimate: _____

$$\begin{array}{r} 726 \\ -\ 148 \\ \hline \end{array}$$

12. Estimate: _____

$$\begin{array}{r} 543 \\ -\ 358 \\ \hline \end{array}$$

Problem Solving  REAL WORLD

13. Mrs. Cohen has 427 buttons. She uses 195 buttons to make puppets. How many buttons does Mrs. Cohen have left?

_____

14. There were 625 ears of corn and 247 tomatoes sold at a farm stand. How many more ears of corn were sold than tomatoes?

_____

© Houghton Mifflin Harcourt Publishing Company

Lesson Check

1. On Saturday, 453 people go to a school play. On Sunday, 294 people go to the play. How many more people go to the play on Saturday?

Ⓐ 159
Ⓑ 169
Ⓒ 259
Ⓓ 747

2. Corey has 510 marbles. He fills one jar with 165 marbles. How many of Corey's marbles are NOT in the jar?

Ⓐ 675
Ⓑ 455
Ⓒ 350
Ⓓ 345

Spiral Review

3. Pattie brought 64 peppers to sell at the farmers' market. There were 12 peppers left at the end of the day. How many peppers did Pattie sell? (Lesson 1.9)

Ⓐ 50
Ⓑ 52
Ⓒ 62
Ⓓ 78

4. An airplane flies 617 miles in the morning. Then it flies 385 miles in the afternoon. About how many more miles does the airplane fly in the morning? (Lesson 1.8)

Ⓐ about 100 miles
Ⓑ about 200 miles
Ⓒ about 300 miles
Ⓓ about 900 miles

5. What is the unknown number?
(Lesson 1.5)

$$(\blacksquare + 4) + 59 = 70$$

Ⓐ 4
Ⓑ 6
Ⓒ 7
Ⓓ 8

6. Dexter has 128 shells. He needs 283 more shells for his art project. How many shells will Dexter use for his art project? (Lesson 1.6)

Ⓐ 155
Ⓑ 165
Ⓒ 401
Ⓓ 411

© Houghton Mifflin Harcourt Publishing Company

Name _____

Combine Place Values to Subtract

Estimate. Then find the difference.

1. Estimate: **200**

476
− 269

2. Estimate: _____

615
− 342

3. Estimate: _____

508
− 113

4. Estimate: _____

716
− 229

5. Estimate: _____

700
− 326

6. Estimate: _____

325
− 179

7. Estimate: _____

935
− 813

8. Estimate: _____

358
− 292

9. Estimate: _____

826
− 617

10. Estimate: _____

900
− 158

11. Estimate: _____

607
− 568

12. Estimate: _____

973
− 869

Problem Solving  REAL WORLD

13. Bev scored 540 points. This was 158 points more than Ike scored. How many points did Ike score?

_____

14. A youth group earned $285 washing cars. The group's expenses were $79. How much profit did the group make washing cars?

_____

© Houghton Mifflin Harcourt Publishing Company

Lesson Check

1. A television program lasts for 120 minutes. Of that time, 36 minutes are taken up by commercials. What is the length of the actual program without the commercials?

   (A) 84 minutes   (C) 104 minutes

   (B) 94 minutes   (D) 156 minutes

2. Syd spent 215 minutes at the library. Of that time, he spent 120 minutes on the computer. How much of his time at the library did Sid NOT spend on the computer?

   (A) 85 minutes   (C) 105 minutes

   (B) 95 minutes   (D) 335 minutes

Spiral Review

3. Xavier's older brother has 568 songs on his music player. To the nearest hundred, about how many songs are on the music player? (Lesson 1.2)

   (A) 500

   (B) 600

   (C) 700

   (D) 800

4. The students traveled to the zoo in 3 buses. One bus had 47 students. The second bus had 38 students. The third bus had 43 students. How many students in all were on the three buses? (Lesson 1.5)

   (A) 108

   (B) 118

   (C) 128

   (D) 138

5. Callie has 83 postcards in her collection. Of the postcards, 24 are from Canada. The rest of the postcards are from the United States. How many of the postcards are from the United States? (Lesson 1.9)

   (A) 58

   (B) 59

   (C) 61

   (D) 69

6. There were 475 seats set up for the school play. At one performance, 189 of the seats were empty. How many seats were filled at that performance? (Lesson 1.10)

   (A) 286

   (B) 296

   (C) 314

   (D) 396

© Houghton Mifflin Harcourt Publishing Company

Name _____

Problem Solving • Model Addition and Subtraction

Use the bar model to solve the problem.

1. Elena went bowling. Elena's score in the first game was 127. She scored 16 more points in the second game than in the first game. What was her total score?

127	16

▲ points

$127 + 16 = ▲$

$143 = ▲$

127	143

■ points

$127 + 143 = ■$

$270 = ■$

270 points

---

2. Mike's Music sold 287 CDs on the first day of a 2-day sale. The store sold 96 more CDs on the second day than on the first day. How many CDs in all were sold during the 2-day sale?

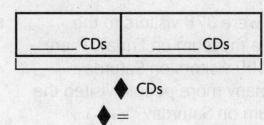

_____ CDs	_____ CDs

★ CDs

★ =

_____ CDs	_____ CDs

◆ CDs

◆ =

_____

© Houghton Mifflin Harcourt Publishing Company

Lesson Check

1. Ms. Hinely picked 46 tomatoes from her garden on Friday. On Saturday, she picked 17 tomatoes. How many tomatoes did she pick in all?

Ⓐ 109   Ⓒ 53

Ⓑ 63   Ⓓ 29

2. Rosa read 57 pages of a book in the morning. She read 13 fewer pages in the afternoon. How many pages did Rosa read in the afternoon?

Ⓐ 44   Ⓒ 70

Ⓑ 60   Ⓓ 83

Spiral Review

3. Mike has 57 action figures. Alex has 186 action figures. Which is the best estimate of the number of action figures Mike and Alex have altogether? (Lesson 1.8)

Ⓐ 150

Ⓑ 250

Ⓒ 350

Ⓓ 400

4. There are 500 sheets of paper in the pack Hannah bought. She has used 137 sheets already. How many sheets of paper does Hannah have left? (Lesson 1.11)

Ⓐ 363

Ⓑ 463

Ⓒ 400

Ⓓ 637

5. There were 378 visitors to the science museum on Friday. There were 409 visitors on Saturday. How many more people visited the museum on Saturday? (Lesson 1.7)

Ⓐ 25

Ⓑ 31

Ⓒ 171

Ⓓ 787

6. Ravi scores 247 points in a video game. How many more points does he need to score a total of 650? (Lesson 1.10)

Ⓐ 897

Ⓑ 430

Ⓒ 417

Ⓓ 403

© Houghton Mifflin Harcourt Publishing Company

Name _____

Chapter 1 Extra Practice

Lesson 1.1

Find the sum. Then use the Commutative Property of Addition to write the related addition sentence.

1. 5 + 7 = ____

____ + ____ = ____

2. 4 + 9 = ____

____ + ____ = ____

3. 0 + 5 = ____

____ + ____ = ____

Lesson 1.2

Round to the nearest ten and hundred.

1. 622 _____

_____

2. 307 _____

_____

3. 867 _____

_____

Lesson 1.3

Use rounding or compatible numbers to estimate the sum.

1.   24   ____
    + 82   + ____

2.   112   ____
    + 279   + ____

3.   583   ____
    + 169   + ____

Lesson 1.4

Use mental math to find the sum.

1. 71 + 99 = _____

2. 38 + 58 = _____

3. 307 + 418 = _____

Lesson 1.5

Use addition properties and strategies to find the sum.

1. 13 + 47 + 21 + 79 = _____

2. 55 + 18 + 15 + 43 = _____

© Houghton Mifflin Harcourt Publishing Company

Lessons 1.6 - 1.7

Estimate. Then find the sum.

1. Estimate: _____
$$
\begin{array}{r}
325 \\
+\ 389 \\
\hline
\end{array}
$$

2. Estimate: _____
$$
\begin{array}{r}
219 \\
+\ 445 \\
\hline
\end{array}
$$

3. Estimate: _____
$$
\begin{array}{r}
437 \\
+\ 146 \\
\hline
\end{array}
$$

4. Estimate: _____
$$
\begin{array}{r}
308 \\
+\ 593 \\
\hline
\end{array}
$$

Lesson 1.8

Use rounding or compatible numbers to estimate the difference.

1.
$$
\begin{array}{r}
82 \\
-\ 44 \\
\hline
\end{array}
$$
_____ − _____

2.
$$
\begin{array}{r}
192 \\
-\ 78 \\
\hline
\end{array}
$$
_____ − _____

3.
$$
\begin{array}{r}
618 \\
-\ 369 \\
\hline
\end{array}
$$
_____ − _____

Lesson 1.9

Use mental math to find the difference.

1. $92 - 41 =$ _____

2. $451 - 125 =$ _____

3. $703 - 359 =$ _____

Lessons 1.10 - 1.11

Estimate. Then find the difference.

1. Estimate: _____
$$
\begin{array}{r}
622 \\
-\ 354 \\
\hline
\end{array}
$$

2. Estimate: _____
$$
\begin{array}{r}
506 \\
-\ 189 \\
\hline
\end{array}
$$

3. Estimate: _____
$$
\begin{array}{r}
763 \\
-\ 295 \\
\hline
\end{array}
$$

4. Estimate: _____
$$
\begin{array}{r}
848 \\
-\ 209 \\
\hline
\end{array}
$$

Lesson 1.12

1. Sara read 81 pages in her book. Colin read 64 pages in his book. How many more pages did Sara read than Colin?

2. Herb planted 28 pea plants. He planted 15 fewer tomato plants. How many pea and tomato plants did Herb plant in all?

© Houghton Mifflin Harcourt Publishing Company

School-Home Letter

Dear Family,

During the next few weeks, our math class will learn about interpreting and representing data.

You can expect to see homework that provides practice with tally tables, frequency tables, picture graphs, bar graphs, and line plots.

Here is a sample of how your child will be taught to solve problems using a bar graph.

© Houghton Mifflin Harcourt Publishing Company

Vocabulary

bar graph A graph that uses bars to show data

data Information that is collected about people or things

frequency table A frequency table uses numbers to record data.

line plot A line plot uses marks to record each piece of data above a number line.

picture graph A picture graph uses small pictures or symbols to show information.

🔑 MODEL   Use a Bar Graph to Solve a Problem

Use the bar graph. How many more sports books than nature books does Richard have?

STEP 1

Identify the bars for Sports and Nature.

STEP 2

Count along the scale to find the difference between the bars. The difference is 5 books.

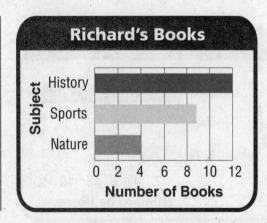

So, Richard has 5 more sports books than nature books.

Tips

Reading Scales

To make reading the length or height of a bar easier, use a straightedge or ruler to line up one end of the bar with the number on the scale.

Activity

Look for bar graphs in magazines and newspapers or help your child create his or her own bar graphs. Then ask questions such as "how many more" and "how many fewer" and help your child find the answers.

Carta
para la casa

Estimada familia,

Durante la próximas semanas, en la clase de matemáticas aprenderemos acerca de interpretar y representar problemas usando una gráfica de barras datos.

Llevaré a la casa tareas que sirven para poner en práctica las tablas de frecuencia, las gráficas de dibujos, las gráficas de barras y los diagramas de puntos.

Este es un ejemplo de la manera como aprenderemos a resolver problemas usando una gráfica de barras .

Vocabulario

gráfica de barras Una gráfica que muestra los datos por medio de barras

datos La información que se recolecta sobre las personas o cosas

tabla de frecuencia Una tabla de frecuencia registra los datos por medio de números.

diagrama de puntos Un diagrama de puntos usa marcas para anotar cada pieza de datos en una recta numérica.

gráfica de dibujos Una gráfica de dibujos muestra la información por medio de dibujos pequeños o símbolos.

🔑 MODELO Usar una gráfica de barras para resolver un problema

Usa la gráfica de barras. ¿Cuántos libros más de deportes que de la naturaleza tiene Richard?

PASO 1

Identifica las barras para Deportes y Naturaleza.

PASO 2

Cuenta a lo largo de la escala para hallar la diferencia entre las barras. La diferencia es 5 libros.

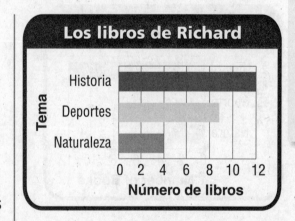

Los libros de Richard

Tema: Historia, Deportes, Naturaleza

Número de libros: 0 2 4 6 8 10 12

Pistas

Escalas

Para leer más fácil la longitud o altura de una barra, usa una orilla recta o una regla para alinear un extremo de la barra con el número de la escala.

Entonces, Richard tiene 5 libros más de deportes que de la naturaleza.

Actividad

Busque y recorte gráficas de barras de revistas o periódicos o ayude a su hijo a crear sus propias gráficas de barras. Después haga preguntas como "cuántos más" y "cuántos menos". Ayúdelo a hallar las respuestas.

© Houghton Mifflin Harcourt Publishing Company

Name _____

Problem Solving • Organize Data

Use the Favorite School Subject tables for 1–4.

1. The students in two third-grade classes recorded their favorite school subject. The data are in the tally table. How many fewer students chose science than chose social studies as their favorite school subject?

   Think: Use the data in the tally table to record the data in the frequency table. Then solve the problem.

   social studies: __12__ students

   science: __5__ students

   $12 - 5 =$ __7__

   So, __7__ fewer students chose science.

2. What subject did the least number of students choose?

   _____

3. How many more students chose math than language arts as their favorite subject?

   _____ more students

4. Suppose 3 students changed their vote from math to science. Describe how the frequency table would change.

   _____

   _____

   _____

   _____

   _____

Favorite School Subject	
Subject	Tally
Math	卌 卌 I
Science	卌
Language Arts	卌 II
Reading	卌 IIII
Social Studies	卌 卌 II

Favorite School Subject	
Subject	Number
Math	
Science	5
Language Arts	
Reading	
Social Studies	12

© Houghton Mifflin Harcourt Publishing Company

Lesson Check

The tally table shows the cards in Kyle's sports card collection.

1. How many hockey and football cards does Kyle have combined?

   Ⓐ 5
   Ⓑ 8
   Ⓒ 12
   Ⓓ 13

Kyle's Sports Cards	
Sport	Tally
Baseball	卌 IIII
Hockey	卌
Basketball	III
Football	卌 III

Spiral Review

2. There are 472 people in the concert hall. What is 472 rounded to the nearest hundred? (Lesson 1.2)

   Ⓐ 400
   Ⓑ 470
   Ⓒ 500
   Ⓓ 600

3. Max and Anna played a video game as a team. Max scored 463 points and Anna scored 329 points. How many points did they score in all? (Lesson 1.12)

   Ⓐ 892
   Ⓑ 792
   Ⓒ 782
   Ⓓ 134

4. Judy has 573 baseball cards in her collection. Todd has 489 baseball cards in his collection. How many fewer cards does Todd have than Judy? (Lesson 1.10)

   Ⓐ 84
   Ⓑ 94
   Ⓒ 116
   Ⓓ 184

5. Ms. Westin drove 542 miles last week and 378 miles this week on business. How many miles in all did she drive on business during the two weeks? (Lesson 1.7)

   Ⓐ 810 miles
   Ⓑ 820 miles
   Ⓒ 910 miles
   Ⓓ 920 miles

© Houghton Mifflin Harcourt Publishing Company

Name _____

Use Picture Graphs

Use the Math Test Scores picture graph for 1–7.

Mrs. Perez made a picture graph of her students' scores on a math test.

Math Test Scores	
100	★★★★★
95	★★★
90	★★★★
85	★

Key: Each ★ = 4 students.

1. How many students scored 100? How can you find the answer?

   To find the number of students who scored 100, count each star as 4 students. So, 20 students scored 100.

2. What does stand for?

   _____
   _____

3. How many students in all scored 100 or 95?

   _____

4. How many more students scored 90 than 85?

   _____

5. How many students in all took the test?

   _____

Problem Solving REAL WORLD

6. Suppose the students who scored 85 and 90 on the math test take the test again and score 95. How many stars would you have to add to the picture graph next to 95?

   _____

7. If 2 more students took the math test and both made a score of 80, what would the picture graph look like?

   _____
   _____

© Houghton Mifflin Harcourt Publishing Company

Lesson Check

1. Karen asked her friends to name their favorite type of dog.

Favorite Dog	
Retriever	🦴 🦴 🦴 🦴 🦴 🦴
Poodle	🦴 🦴 🦴
Terrier	🦴 🦴

Key: Each 🦴 = 2 people.

How many people chose poodles?

Ⓐ 10   Ⓒ 4

Ⓑ 6   Ⓓ 3

2. Henry made a picture graph to show what topping people like on their pizza. This is his key.

Each = 6 people.

What does 🍕 🍕 stand for?

Ⓐ 2 people

Ⓑ 6 people

Ⓒ 9 people

Ⓓ 12 people

Spiral Review

3. Estimate the sum. (Lesson 1.3)

$$\begin{array}{r} 523 \\ + \ 295 \\ \hline \end{array}$$

Ⓐ 900   Ⓒ 700

Ⓑ 800   Ⓓ 600

4. Estimate the difference. (Lesson 1,8)

$$\begin{array}{r} 610 \\ - \ 187 \\ \hline \end{array}$$

Ⓐ 800   Ⓒ 500

Ⓑ 600   Ⓓ 400

5. What is 871 rounded to the nearest ten? (Lesson 1.2)

Ⓐ 900

Ⓑ 880

Ⓒ 870

Ⓓ 800

6. What is 473 rounded to the nearest hundred? (Lesson 1.2)

Ⓐ 400

Ⓑ 470

Ⓒ 500

Ⓓ 570

© Houghton Mifflin Harcourt Publishing Company

Name _____

Make Picture Graphs

Ben asked his classmates about their favorite kind of TV show. He recorded their responses in a frequency table. Use the data in the table to make a picture graph.

Favorite TV Show	
Type	Number
Cartoons	9
Sports	6
Movies	3

Follow the steps to make a picture graph.

Step 1 Write the title at the top of the graph.

Step 2 Look at the numbers in the table. Tell how many students each picture represents for the key.

Step 3 Draw the correct number of pictures for each type of show.

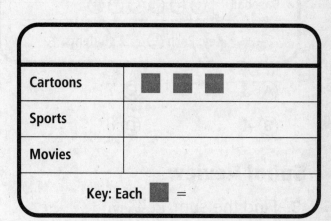

Use your picture graph for 1–5.

1. What title did you give the graph?

   _____

   _____

2. What key did you use?

   _____

   _____

3. How many pictures did you use to represent sports?

   _____

Problem Solving REAL WORLD

4. How many pictures would you draw if 12 students chose game shows as their favorite kind of TV show?

   _____

5. What key would you use if 10 students chose cartoons?

   _____

© Houghton Mifflin Harcourt Publishing Company

Lesson Check

1. Sandy made a picture graph to show the sports her classmates like to play. How many fewer students chose baseball than chose soccer?

Favorite Sport	
Basketball	○○○○○○○
Soccer	○○○○○○○○(
Baseball	○○○○○

Key: Each ○ = 2 students.

Ⓐ 3   Ⓒ 7

Ⓑ 4   Ⓓ 8

2. Tommy is making a picture graph to show his friends' favorite kind of music. He plans to use one musical note to represent 2 people. How many notes will he use to represent that 4 people chose country music?

Ⓐ 2

Ⓑ 4

Ⓒ 6

Ⓓ 8

Spiral Review

3. Find the sum. (Lesson 1.7)

$$490 \\ + 234$$

Ⓐ 256   Ⓒ 664

Ⓑ 624   Ⓓ 724

4. Sophie wrote odd numbers on her paper. Which number was NOT a number that Sophie wrote? (Lesson 1.1)

Ⓐ 5   Ⓒ 13

Ⓑ 11   Ⓓ 20

5. Miles ordered 126 books to give away at the store opening. What is 126 rounded to the nearest hundred? (Lesson 1.2)

Ⓐ 230

Ⓑ 200

Ⓒ 130

Ⓓ 100

6. Estimate the difference. (Lesson 1.8)

$$422 \\ - 284$$

Ⓐ 100

Ⓑ 180

Ⓒ 200

Ⓓ 700

© Houghton Mifflin Harcourt Publishing Company

Name _____

Use Bar Graphs

Use the After-Dinner Activities bar graph for 1–6.

The third-grade students at Case Elementary School were asked what they spent the most time doing last week after dinner. The results are shown in the bar graph at the right.

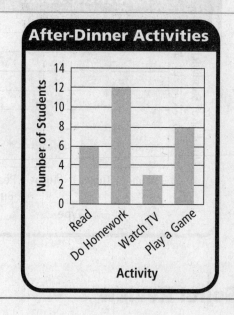

1. How many students spent the most time watching TV after dinner?

   3 students

2. How many students in all answered the survey?

   _____

3. How many students in all played a game or read?

   _____

4. How many fewer students read than did homework?

   _____

5. How many more students read than watched TV?

   _____

Problem Solving REAL WORLD

6. Suppose 3 students changed their answers to reading instead of doing homework. Where would the bar for reading end?

   _____

© Houghton Mifflin Harcourt Publishing Company

Lesson Check

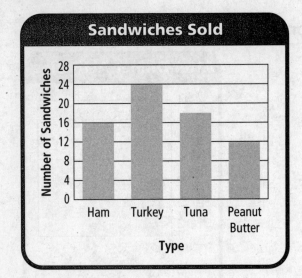

Sandwiches Sold

1. The bar graph shows the number of sandwiches sold at Lisa's sandwich cart yesterday. How many tuna sandwiches were sold?

   (A) 12

   (B) 16

   (C) 18

   (D) 20

Spiral Review

2. What is 582 rounded to the nearest ten? (Lesson 1.2)

   (A) 500

   (B) 580

   (C) 590

   (D) 600

3. Savannah read 178 minutes last week. What is 178 rounded to the nearest hundred? (Lesson 1.2)

   (A) 400   (C) 200

   (B) 280   (D) 180

4. Estimate the difference. (Lesson 1.8)

   $$371$$
   $$-\ \ 99$$

   (A) 500   (C) 300

   (B) 400   (D) 200

5. Estimate the difference. (Lesson 1.8)

   $$625$$
   $$-\ 248$$

   (A) 800   (C) 400

   (B) 500   (D) 300

© Houghton Mifflin Harcourt Publishing Company

Make Bar Graphs

Ben asked some friends to name their favorite breakfast food. He recorded their choices in the frequency table at the right.

Favorite Breakfast Food	
Food	Number of Votes
Waffles	8
Cereal	14
Pancakes	12
Oatmeal	4

1. Complete the bar graph by using Ben's data.

Favorite Breakfast Food

Use your bar graph for 2–5.

2. Which food did the most people choose as their favorite breakfast food?

   _____

3. How many people chose waffles as their favorite breakfast food?

   _____

4. How did you know how high to draw the bar for pancakes?

   _____

   _____

   _____

5. Suppose 6 people chose oatmeal as their favorite breakfast food. How would you change the bar graph?

   _____

   _____

   _____

   _____

© Houghton Mifflin Harcourt Publishing Company

Lesson Check

Favorite Pizza Topping

1. Gary asked his friends to name their favorite pizza topping. He recorded the results in a bar graph. How many people chose pepperoni?

   (A) 6   (C) 4

   (B) 5   (D) 1

2. Suppose 3 more friends chose mushrooms. Where would the bar for mushrooms end?

   (A) 2   (C) 6

   (B) 4   (D) 8

Spiral Review

3. Estimate the sum. (Lesson 1.3)

   458
   + 214

   (A) 700   (C) 300

   (B) 600   (D) 200

4. Matt added $14 + 0$. What is the correct sum? (Lesson 1.1)

   (A) 140   (C) 1

   (B) 14   (D) 0

5. There are 682 runners registered for an upcoming race. What is 682 rounded to the nearest hundred? (Lesson 1.2)

   (A) 600

   (B) 680

   (C) 700

   (D) 780

6. There are 187 new students this year at Maple Elementary. What is 187 rounded to the nearest ten? (Lesson 1.2)

   (A) 100

   (B) 180

   (C) 190

   (D) 200

© Houghton Mifflin Harcourt Publishing Company

Name _____

Solve Problems Using Data

Use the Favorite Hot Lunch bar graph for 1–3.

1. How many more students chose pizza than chose grilled cheese?

   Think: Subtract the number of students who chose grilled cheese, 2, from the number of students who chose pizza, 11.

   $11 - 2 = 9$ _____ more students

2. How many students did not choose chicken patty?   _____ students

3. How many fewer students chose grilled cheese than chose hot dog?

   _____ fewer students

Use the Ways to Get to School bar graph for 4–7.

4. How many more students walk than ride in a car to get to school?

   _____ more students

5. How many students walk and ride a bike combined?

   _____ students

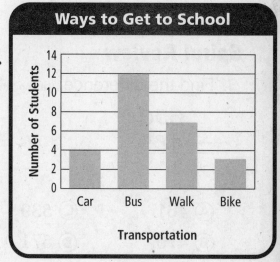

Problem Solving REAL WORLD

6. Is the number of students who get to school by car and bus greater than or less than the number of students who get to school by walking and biking? **Explain.**

   _____

   _____

7. **What if** 5 more students respond that they get to school by biking? Would more students walk or ride a bike to school? **Explain.**

   _____

   _____

© Houghton Mifflin Harcourt Publishing Company

Lesson Check

1. How many fewer votes were for bench repair than for food drive?

   (A) 9

   (B) 10

   (C) 11

   (D) 16

2. How many votes were there in all?

   (A) 4   (C) 32

   (B) 14   (D) 34

Community Project

Food Drive

Wall Mural

Bench Repair

Park Pick Up

Project

0 2 4 6 8 10 12 14
Number of Votes

Spiral Review

3. Find the difference. (Lesson 1.10)

   650
   − 189

   (A) 461   (C) 539

   (B) 479   (D) 571

4. Greyson has 75 basketball cards. What is 75 rounded to the nearest ten? (Lesson 1.2)

   (A) 60

   (B) 70

   (C) 80

   (D) 90

5. Sue spent $18 on a shirt, $39 on a jacket, and $12 on a hat. How much did she spend in all? (Lesson 1.5)

   (A) $79   (C) $57

   (B) $69   (D) $51

6. There are 219 adults and 174 children at a ballet. How many people are at the ballet in all?

   (Lesson 1.7)

   (A) 45   (C) 383

   (B) 293   (D) 393

© Houghton Mifflin Harcourt Publishing Company

Use and Make Line Plots

Use the data in the table to make a line plot.

How Many Shirts Were Sold at Each Price?	
Price	Number Sold
$11	1
$12	4
$13	6
$14	4
$15	0
$16	2

```
    |    |    |    |    |    |
  $11  $12  $13  $14  $15  $16
```

How Many Shirts Were Sold at Each Price?

1. How many shirts sold for $12?

4 shirts
_____

2. At which price were the most shirts sold?

_____

3. How many shirts in all were sold?

_____

4. How many shirts were sold for $13 or more?

_____

Problem Solving REAL WORLD

Use the line plot above for 5–6.

5. Were more shirts sold for less than $13 or more than $13? **Explain.**

_____

_____

6. Is there any price for which there are no data? **Explain.**

_____

_____

© Houghton Mifflin Harcourt Publishing Company

Lesson Check

1. Pedro made a line plot to show the heights of the plants in his garden. How many plants are less than 3 inches tall?

   Ⓐ 4   Ⓒ 10

   Ⓑ 5   Ⓓ 16

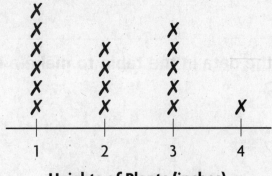

Heights of Plants (inches)

Spiral Review

2. Find the sum. (Lesson 1.7)

   $$642$$
   $$+\ 259$$

   Ⓐ 383
   Ⓑ 801
   Ⓒ 891
   Ⓓ 901

3. Find the difference. (Lesson 1.10)

   $$460$$
   $$-\ 309$$

   Ⓐ 61
   Ⓑ 151
   Ⓒ 161
   Ⓓ 169

4. There were 262 hamburgers cooked for the school fair. What is 262 rounded to the nearest hundred? (Lesson 1.2)

   Ⓐ 200
   Ⓑ 260
   Ⓒ 270
   Ⓓ 300

5. Makenzie has 517 stickers in her collection. What is 517 rounded to the nearest ten? (Lesson 1.2)

   Ⓐ 500
   Ⓑ 510
   Ⓒ 520
   Ⓓ 600

© Houghton Mifflin Harcourt Publishing Company

Name _____

Chapter 2 Extra Practice

Lesson 2.1

Use the Pets tables for 1–2.

1. Manny collected data about pets owned by students in his class. Complete Manny's tally table and frequency table.

Pets			
Pets	Tally		
Cat			4
Dog			2
Bird			1
Fish			1

2. How many more students have cats than have dogs and birds combined?

_____

Lessons 2.2 - 2.3

Use the Seashells picture graph for 1–3.

1. Maggie has a picture graph that shows the seashells she collected. How many seashells did Maggie collect in all?

_____

2. How many more cockle shells did she collect than lightning whelks?

_____

3. **What if** the key were "Each 🐚 = 5 shells?" How many pictures would there be for conch?

_____

Seashells

Cockle	🐚 🐚 🐚 🐚
Conch	🐚 🐚
Lightning Whelk	🐚 🐚 🐚

Key: Each 🐚 = 10 shells.

© Houghton Mifflin Harcourt Publishing Company

Lessons 2.4 - 2.6

Use the Bicycle Rides frequency table for 1–3.

Bicycle Rides	
Day	Number of Miles
Monday	4
Wednesday	9
Saturday	12

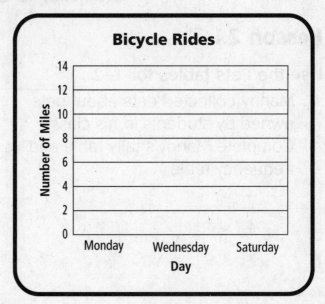

Bicycle Rides

1. The frequency table shows the number of miles Sean rode on his bicycle. Use the data in the frequency table to complete the bar graph.

2. How many more miles did Sean ride on Saturday than on Monday?

_____

3. Write a number sentence to show how many miles in all Sean rode on his bicycle.

_____

Lesson 2.7

Use the Number of Beads line plot for 1–3.

1. Kim is making bead necklaces. She records the number of beads on the different necklaces on a line plot. How many necklaces have exactly 50 beads?

_____

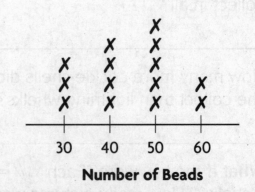

Number of Beads

2. How many necklaces have fewer than 40 beads?

_____

3. How many necklaces have 50 or more beads?

_____

© Houghton Mifflin Harcourt Publishing Company

Vocabulary

array A set of objects in rows and columns

equal groups Groups that have the same number of objects

factor A number that is multiplied by another number to find a product

multiply When you multiply, you combine equal groups to find how many in all.

product The answer in a multiplication problem

Dear Family,

During the next few weeks, our math class will be learning about multiplication. We will learn how addition is related to multiplication and how to multiply with the factors 0 and 1.

You can expect to see homework that provides practice with multiplication.

Here is a sample of how your child will be shown the relationship between addition and multiplication.

🔑 MODEL Relate Addition and Multiplication

This is how we will add or multiply to solve problems about equal groups.

Add.

STEP 1

Draw 2 counters in each rectangle to show 4 equal groups.

STEP 2

Write an addition sentence to find how many counters in all.

$2 + 2 + 2 + 2 = 8$

Multiply.

STEP 1

Draw 2 counters in each rectangle to show 4 equal groups.

STEP 2

Write a multiplication sentence to find how many counters in all.

$4 \times 2 = 8$

Tips

Skip Counting

Skip counting is another way to count equal groups to find how many in all. For example, there are 4 groups with 2 counters in each group, so skip counting by 2s can be used: 2, 4, 6, 8. There are 8 counters in all.

Activity

Help your child arrange 3 equal groups of like objects (no more than 10 objects in each group). Then have him or her write an addition sentence and a multiplication sentence to find how many objects in all.

© Houghton Mifflin Harcourt Publishing Company

Carta para la casa

© Houghton Mifflin Harcourt Publishing Company

Vocabulario

arreglo Un grupo de objetos organizados en filas y columnas

grupos iguales Grupos que tienen la misma cantidad de objetos

factor Un número que se multiplica por otro número para hallar el producto

multiplicar Cuando uno multiplica, combina grupos iguales para hallar cuántos hay en total.

producto El resultado de una multiplicación

Querída Familia,

Durante las próximas semanas, en la clase de matemáticas aprenderemos sobre la multiplicación. Aprenderemos cómo la suma se relaciona con la multiplicación y a multiplicar por los factores 0 y 1.

Llevaré a la casa tareas que sirven para practicar la multiplicación.

Este es un ejemplo de la manera como aprenderemos la relación entre la suma y la multiplicación.

🔑 MODELO Relacionar la suma y multiplicación

Así es como vamos a sumar o multiplicar para resolver problemas de grupos iguales.

Suma.

PASO 1

Dibuja 2 fichas en cada rectángulo para mostrar 4 grupos iguales.

PASO 2

Escribe un enunciado de suma para hallar cuántas fichas hay en total.

$2 + 2 + 2 + 2 = 8$

Multiplica.

PASO 1

Dibuja 2 fichas en cada rectángulo para mostrar 4 grupos iguales.

PASO 2

Escribe un enunciado de multiplicación para hallar cuántas fichas hay en total.

$4 \times 2 = 8$

Pistas

Contar salteado
Contar salteado es otra manera de contar grupos iguales para hallar cuánto hay en total. Por ejemplo, hay 4 grupos con 2 fichas cada uno, por lo tanto puedes contar salteado de 2 en 2: 2, 4, 6, 8. Hay 8 fichas en total.

Actividad

Ayude a su hijo a formar 3 grupos iguales de objetos parecidos (no más de 10 objetos en cada grupo). Después, pídale que escriba un enunciado de suma y uno de multiplicación para hallar cuántos objetos hay en total.

Name _____

Count Equal Groups

Draw equal groups. Skip count to find how many.

1. 2 groups of 2 ____4____

2. 3 groups of 6 _____

3. 5 groups of 3 _____

4. 4 groups of 5 _____

Count equal groups to find how many.

5.

_____ groups of _____

_____ in all

6.

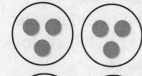

_____ groups of _____

_____ in all

Problem Solving REAL WORLD

7. Marcia puts 2 slices of cheese on each sandwich. She makes 4 cheese sandwiches. How many slices of cheese does Marcia use in all?

_____

8. Tomas works in a cafeteria kitchen. He puts 3 cherry tomatoes on each of 5 salads. How many tomatoes does he use?

_____

© Houghton Mifflin Harcourt Publishing Company

Lesson Check

1. Jen makes 3 bracelets. Each bracelet has 3 beads. How many beads does Jen use?

   (A) 12   (C) 6

   (B) 9   (D) 3

2. Ian has 5 cards to mail. Each card needs 2 stamps. How many stamps does Ian need?

   (A) 2   (C) 10

   (B) 5   (D) 15

Spiral Review

3. There were 384 people at a play on Friday night. There were 512 people at the play on Saturday night. Which is the best estimate of the total number of people who attended the play on both nights? (Lesson 1.3)

   (A) 900   (C) 700

   (B) 800   (D) 500

4. Walking the Dog Pet Store has 438 leashes in stock. They sell 79 leashes during a one-day sale. How many leashes are left in stock after the sale? (Lesson 1.10)

   (A) 459   (C) 369

   (B) 441   (D) 359

5. The Lakeside Tour bus traveled 490 miles on Saturday and 225 miles on Sunday. About how many more miles did it travel on Saturday? (Lesson 1.8)

   (A) 500 miles   (C) 300 miles

   (B) 400 miles   (D) 100 miles

6. During one week at Jackson School, 210 students buy milk and 196 students buy juice. How many drinks are sold that week? (Lesson 1.7)

   (A) 496   (C) 396

   (B) 406   (D) 306

© Houghton Mifflin Harcourt Publishing Company

Relate Addition and Multiplication

Draw a quick picture to show the equal groups. Then write related addition and multiplication sentences.

1. 3 groups of 5

<u>5</u> + <u>5</u> + <u>5</u> = <u>15</u>

<u>3</u> × <u>5</u> = <u>15</u>

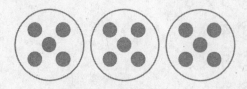

2. 3 groups of 4

___ + ___ + ___ = ___

___ × ___ = ___

3. 4 groups of 3

___ + ___ + ___ + ___ = ___

___ × ___ = ___

4. 5 groups of 2

___ + ___ + ___ + ___ + ___ = ___

___ × ___ = ___

Complete. Write a multiplication sentence.

5. 7 + 7 + 7 = ___

___ × ___ = ___

6. 3 + 3 + 3 = ___

___ × ___ = ___

Problem Solving REAL WORLD

7. There are 6 jars of pickles in a box. Ed has 3 boxes of pickles. How many jars of pickles does he have in all? Write a multiplication sentence to find the answer.

___ × ___ = ___ jars

8. Each day, Jani rides her bike 5 miles. How many miles does Jani ride in all in 4 days? Write a multiplication sentence to find the answer.

___ × ___ = ___ miles

© Houghton Mifflin Harcourt Publishing Company

Lesson Check

1. Which is another way to show

   $3 + 3 + 3 + 3 + 3 + 3$?

   (A) 5×3

   (B) 4×3

   (C) 8×3

   (D) 6×3

2. Use the model. How many counters are there in all?

   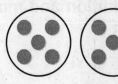

   (A) 8   (C) 12

   (B) 10   (D) 14

Spiral Review

3. A school gave 884 pencils to students on the first day of school. What is 884 rounded to the nearest hundred? (Lesson 1.2)

   (A) 800   (C) 890

   (B) 880   (D) 900

4. Find the difference. (Lesson 1.10)

   $$632 - 274$$

   (A) 906   (C) 358

   (B) 442   (D) 354

5. The line plot below shows how many points Trevor scored in 20 games. (Lesson 2.7)

   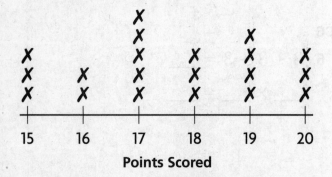

   Points Scored

   In how many games did Trevor score at least 18 points?

   (A) 3   (C) 6

   (B) 5   (D) 10

6. Darrien read 97 pages last week. Evan read 84 pages last week. How many pages in all did the boys read? (Lesson 1.7)

   (A) 13

   (B) 171

   (C) 181

   (D) 271

© Houghton Mifflin Harcourt Publishing Company

Name _____

Skip Count on a Number Line

Draw jumps on the number line to show equal groups. Find the product.

1. 6 groups of 3

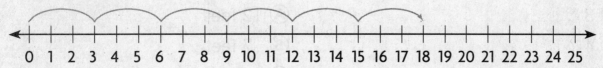

$6 \times 3 =$ ___**18**___

2. 3 groups of 5

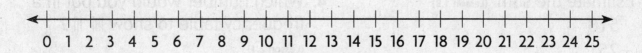

$3 \times 5 =$ _____

Write the multiplication sentence the number line shows.

3. 2 groups of 6

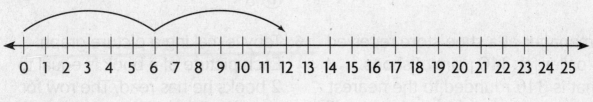

_____ $\times$ _____ = _____

Problem Solving REAL WORLD

4. Allie is baking muffins for students in her class. There are 6 muffins in each baking tray. She bakes 5 trays of muffins. How many muffins is she baking in all?

_____

5. A snack package has 4 cheese sticks. How many cheese sticks are in 4 packages?

_____

© Houghton Mifflin Harcourt Publishing Company

Lesson Check

1. Louise skip counts by 4 on a number line to find 5×4. How many jumps should she draw on the number line?

- (A) 3
- (B) 4
- (C) 5
- (D) 9

2. Theo needs 4 boards that are each 3 feet long to make bookshelves. How many feet of boards does he need altogether?

- (A) 12 feet
- (B) 7 feet
- (C) 4 feet
- (D) 3 feet

Spiral Review

3. Estimate the sum. (Lesson 1.3)

$$518$$
$$+251$$

- (A) 200
- (C) 800
- (B) 700
- (D) 900

4. Which number would you put in a frequency table to show 卌 ‖‖? (Lesson 2.1)

- (A) 5
- (B) 6
- (C) 7
- (D) 8

5. A manager at a shoe store received an order for 346 pairs of shoes. What is 346 rounded to the nearest hundred? (Lesson 1.2)

- (A) 400
- (B) 350
- (C) 340
- (D) 300

6. Toby is making a picture graph. Each picture of a book is equal to 2 books he has read. The row for Month 1 has 3 pictures of books. How many books did Toby read during Month 1? (Lesson 2.2)

- (A) 2
- (B) 3
- (C) 6
- (D) 8

© Houghton Mifflin Harcourt Publishing Company

Name _____

Problem Solving •
Model Multiplication
Draw a diagram to solve each problem.

1. Robert put some toy blocks into 3 rows.
   There are 5 blocks in each row. How
   many blocks are there in all?

   _____ **15 blocks** _____

2. Mr. Fernandez is putting tiles on his
   kitchen floor. There are 2 rows with
   9 tiles in each row. How many tiles are
   there in all?

   _____

3. In Jillian's garden, there are 3 rows of carrots,
   2 rows of string beans, and 1 row of peas.
   There are 8 plants in each row. How many
   plants are there in all?

   _____

4. In Sorhab's classroom, there are 3 rows
   with 7 desks in each row. How many
   desks are there in all?

   _____

5. Maya visits the movie rental store. On one
   wall, there are 6 DVDs on each of 5 shelves.
   On another wall, there are 4 DVDs on each of
   4 shelves. How many DVDs are there in all?

   _____

6. The media center at Josh's school has a
   computer area. The first 4 rows have
   6 computers each. The fifth row has 4 computers.
   How many computers are there in all?

   _____

© Houghton Mifflin Harcourt Publishing Company

Lesson Check

1. There are 5 shelves of video games in a video store. There are 6 video games on each shelf. How many video games are there in all?

   (A) 35
   (B) 30
   (C) 20
   (D) 11

2. Ken watches a marching band. He sees 2 rows of flute players. Six people are in each row. He sees 8 trombone players. How many flute or trombone players does Ken see?

   (A) 2
   (B) 6
   (C) 16
   (D) 20

Spiral Review

3. What is the sum of 438 and 382? (Lesson 1.7)

   (A) 720   (C) 820
   (B) 810   (D) 910

4. Estimate the sum. (Lesson 1.3)

   $$622$$
   $$+ \quad 84$$

   (A) 500   (C) 700
   (B) 600   (D) 800

5. Francine uses 167 silver balloons and 182 gold balloons for her store party. How many silver and gold balloons in all does Francine use? (Lesson 1.7)

   (A) 15
   (B) 345
   (C) 349
   (D) 359

6. Yoshi is making a picture graph. Each picture of a soccer ball stands for two goals he scored for his team. The row for January has 9 soccer balls. How many goals did Yoshi score during January? (Lesson 2.2)

   (A) 18
   (B) 16
   (C) 11
   (D) 9

© Houghton Mifflin Harcourt Publishing Company

Name _____

Model with Arrays

Write a multiplication sentence for the array.

1.

   $3 \times 7 = \underline{\ 21\ }$

2.

   $2 \times 5 = \underline{\qquad}$

Draw an array to find the product.

3. $4 \times 2 = \underline{\qquad}$

4. $4 \times 4 = \underline{\qquad}$

5. $3 \times 2 = \underline{\qquad}$

6. $2 \times 8 = \underline{\qquad}$

Problem Solving  REAL WORLD

7. Lenny is moving tables in the school cafeteria. He places all the tables in a 7×4 array. How many tables are in the cafeteria?

   _____

8. Ms. DiMeo directs the school choir. She has the singers stand in 3 rows. There are 8 singers in each row. How many singers are there in all?

   _____

© Houghton Mifflin Harcourt Publishing Company

Lesson Check

1. What multiplication sentence does this array show?

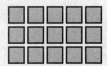

   (A) $2 \times 3 = 6$ (C) $3 \times 4 = 12$

   (B) $6 \times 3 = 18$ (D) $3 \times 5 = 15$

2. What multiplication sentence does this array show?

   (A) $3 \times 9 = 27$ (C) $3 \times 7 = 21$

   (B) $3 \times 8 = 24$ (D) $4 \times 5 = 20$

Spiral Review

3. Use the table to find who traveled 700 miles farther than Paul during summer vacation. (Lesson 1.6)

Summer Vacations

Name	Distance in Miles
Paul	233
Andrew	380
Bonnie	790
Tara	933
Susan	853

   (A) Andrew   (C) Susan

   (B) Bonnie   (D) Tara

4. Use the bar graph to find what hair color most students have. (Lesson 2.4)

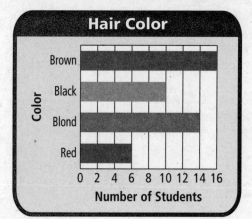

   (A) Brown   (C) Blond

   (B) Black   (D) Red

5. Spencer ordered 235 cans of tomatoes to make salsa for the festival. What is 235 rounded to the nearest ten? (Lesson 1.2)

   (A) 200

   (B) 230

   (C) 240

   (D) 300

6. Which bar would be the longest on a bar graph of the data? (Lesson 2.5)

Favorite Pizza Topping

Topping	Votes
Cheese	5
Pepperoni	4
Vegetable	1
Sausage	3

   (A) Cheese   (C) Vegetable

   (B) Pepperoni   (D) Sausage

© Houghton Mifflin Harcourt Publishing Company

Name _____

Commutative Property of Multiplication

Write a multiplication sentence for the model. Then use the Commutative Property of Multiplication to write a related multiplication sentence.

1. (model)

$$\frac{5}{2} \times \frac{2}{5} = \frac{10}{10}$$

2. (model)

___ × ___ = ___

___ × ___ = ___

3. (model)

___ × ___ = ___

___ × ___ = ___

4. (model)

___ × ___ = ___

___ × ___ = ___

Problem Solving REAL WORLD

5. A garden store sells trays of plants. Each tray holds 2 rows of 8 plants. How many plants are in one tray?

_____

6. Jeff collects toy cars. They are displayed in a case that has 4 rows. There are 6 cars in each row. How many cars does Jeff have?

_____

© Houghton Mifflin Harcourt Publishing Company

Lesson Check

1. Which is an example of the Commutative Property of Multiplication?

(A) $8 \times 4 = 8 \times 4$

(B) $4 \times 2 = 2 \times 4$

(C) $2 \times 8 = 4 \times 4$

(D) $2 + 4 = 2 \times 4$

2. What factor makes the number sentence true?

$7 \times 4 = \blacksquare \times 7$

(A) 2

(B) 4

(C) 7

(D) 28

Spiral Review

3. Ms. Williams drove 149 miles on Thursday and 159 miles on Friday. About how many miles did she drive altogether the two days? (Lesson 1.3)

(A) about 150 miles

(B) about 200 miles

(C) about 300 miles

(D) about 400 miles

4. Inez has 699 pennies and 198 nickels. Estimate how many more pennies than nickels she has. (Lesson 1.8)

(A) about 500

(B) about 600

(C) about 700

(D) about 900

5. This year, the parade had 127 floats. That is 34 fewer floats than last year. How many floats were in the parade last year? (Lesson 1.7)

(A) 161

(B) 151

(C) 103

(D) 93

6. Jeremy made a tally table to record how his friends voted for their favorite pet. His table shows 卌 卌 II next to Dog. How many friends voted for dog? (Lesson 2.1)

(A) 6

(B) 8

(C) 10

(D) 12

© Houghton Mifflin Harcourt Publishing Company

Name _____

Multiply with 1 and 0

Find the product.

1. $1 \times 4 =$ __4__   **2.** $0 \times 8 =$ ___   **3.** $0 \times 4 =$ ___   **4.** $1 \times 6 =$ ___

5. $3 \times 0 =$ ___   **6.** $0 \times 9 =$ ___   **7.** $8 \times 1 =$ ___   **8.** $1 \times 2 =$ ___

9. $0 \times 6 =$ ___   **10.** $4 \times 0 =$ ___   **11.** $7 \times 1 =$ ___   **12.** $1 \times 5 =$ ___

13. $3 \times 1 =$ ___   **14.** $0 \times 7 =$ ___   **15.** $1 \times 9 =$ ___   **16.** $5 \times 0 =$ ___

17. $10 \times 1 =$ ___   **18.** $2 \times 0 =$ ___   **19.** $5 \times 1 =$ ___   **20.** $1 \times 0 =$ ___

21. $0 \times 0 =$ ___   **22.** $1 \times 3 =$ ___   **23.** $9 \times 0 =$ ___   **24.** $1 \times 1 =$ ___

Problem Solving REAL WORLD

25. Peter is in the school play. His teacher gave 1 copy of the play to each of 6 students. How many copies of the play did the teacher hand out?

26. There are 4 egg cartons on the table. There are 0 eggs in each carton. How many eggs are there in all?

_____   _____

© Houghton Mifflin Harcourt Publishing Company

Lesson Check

1. There are 0 bicycles in each bicycle rack. If there are 8 bicycle racks, how many bicycles are there in all?

   (A) 80   (C) 1
   (B) 8   (D) 0

2. What is the product?

   $1 \times 0 =$ ___

   (A) 0   (C) 10
   (B) 1   (D) 11

Spiral Review

3. Mr. Ellis drove 197 miles on Monday and 168 miles on Tuesday. How many miles did he drive in all? (Lesson 1.6)

   (A) 29 miles   (C) 365 miles
   (B) 255 miles   (D) 400 miles

4. What multiplication sentence does the array show? (Lesson 3.5)

   (A) $1 \times 6 = 6$
   (B) $3 \times 2 = 6$
   (C) $2 \times 6 = 12$
   (D) $5 + 1 = 6$

Use the bar graph for 5–6.

5. How many cars were washed on Friday and Saturday combined? (Lesson 2.6)

   (A) 55   (C) 90
   (B) 80   (D) 120

6. How many more cars were washed on Saturday than on Sunday? (Lesson 2.6)

   (A) 95   (C) 25
   (B) 30   (D) 15

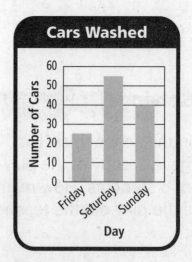

Cars Washed

© Houghton Mifflin Harcourt Publishing Company

Name _____

Chapter 3 Extra Practice

Lesson 3.1

Draw equal groups. Skip count to find how many.

1. 2 groups of 4 _____

2. 4 groups of 3 _____

Lesson 3.2

Draw a quick picture to show the equal groups. Then write related addition and multiplication sentences.

1. 2 groups of 5

2. 3 groups of 2

_____ + _____ = _____   _____ + _____ + _____ = _____

_____ × _____ = _____   _____ × _____ = _____

Lesson 3.3

1. Draw jumps on the number line to show 3 groups of 6.
Find the product.

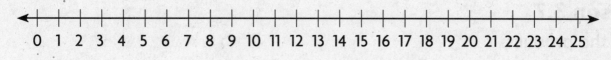

$3 \times 6 =$ _____

2. Write the multiplication sentence the number line shows.

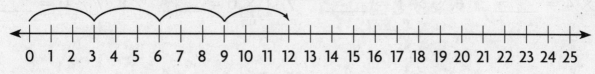

_____ × _____ = _____

© Houghton Mifflin Harcourt Publishing Company

Lesson 3.4

1. Destiny placed her hair ribbons in 3 groups of 5 on her dresser. How many hair ribbons in all does Destiny have? Draw a diagram to solve.

_____

Lesson 3.5

Draw an array to find the product.

1. $2 \times 7 = $ _____

2. $2 \times 6 = $ _____

Lesson 3.6

Write a multiplication sentence for the model. Then use the Commutative Property of Multiplication to write a related multiplication sentence.

1.

2.

_____ × _____ = _____   _____ × _____ = _____

_____ × _____ = _____   _____ × _____ = _____

Lesson 3.7

Find the product.

1. $6 \times 0 = $ _____   2. $5 \times 1 = $ _____   3. $0 \times 9 = $ _____   4. $1 \times 8 = $ _____

5. $1 \times 4 = $ _____   6. $9 \times 1 = $ _____   7. $1 \times 0 = $ _____   8. $7 \times 0 = $ _____

© Houghton Mifflin Harcourt Publishing Company

School-Home Letter

© Houghton Mifflin Harcourt Publishing Company

Dear Family,

During the next few weeks, our math class will be learning how to multiply with the factors 2, 3, 4, 5, 6, 7, 8, 9, and 10.

You can expect to see homework that provides practice with multiplication facts and strategies.

Here is a sample of how your child will be taught to multiply with 3 as a factor.

Vocabulary

Associative Property of Multiplication The property that states that when the grouping of factors is changed, the product remains the same.

Distributive Property The property that states that multiplying a sum by a number is the same as multiplying each addend by the number and then adding the products.

multiple A number that is the product of two counting numbers

🔑 MODEL Multiply with 3

This is one way we will be multiplying with 3 to solve problems.

Teddy made a face on 1 cookie, using 3 raisins. How many raisins will he need for 4 cookies?

Drawing a picture is a way to solve this problem.

 3,   6,   9,   12

Skip count by 3s to find the number of raisins in all.

3, 6, 9, 12

4 groups of 3 is 12. $4 \times 3 = 12$

So, he will need 12 raisins for 4 cookies.

Tips

Another Way to Solve Multiplication Problems

Making an array is another way to solve the problem. Use tiles to make an array of 4 rows with 3 tiles in each row.

Count all the tiles.

4 groups of 3 is 12.
$4 \times 3 = 12$

Activity

Have your child draw more groups of 3 for 5, 6, 7, 8, and 9 cookies. Then have your child answer questions such as "How many raisins would be on 8 cookies? What do you multiply to find out?"

Carta para la casa

Vocabulario

Propiedad asociativa de la multiplicación La propiedad que establece que cuando se cambia la agrupación de los factores, el producto no cambia

Propiedad distributiva La propiedad que establece que multiplicar una suma por un número es lo mismo que multiplicar cada sumando por ese número y luego sumar los productos

múltiplo Un número que es el producto de dos números naturales distintos de cero

Querida familia,

Durante las próximas semanas, en la clase de matemáticas aprenderemos cómo multiplicar con los factores 2, 3, 4, 5, 6, 7, 8, 9 y 10.

Llevaré a la casa tareas que sirven para practicar las operaciones de multiplicación y sus estrategias.

Este es un ejemplo de la manera como aprenderemos a multiplicar por el factor 3.

🔑 MODELO   Multiplicar por 3

Esta es una manera de multiplicar por 3 para resolver problemas.

Teddy hizo una cara en 1 galleta, con 3 pasas.
¿Cuántas pasas necesitará para hacer caras en 4 galletas?

Una manera de resolver el problema es hacer un dibujo.

3,　　　　6,　　　　9,　　　　12

Cuenta salteado de 3 en 3 para hallar el número total de pasas.

3, 6, 9, 12

4 grupos de 3 son 12.　　　$4 \times 3 = 12$

Por tanto, Teddy necesitará 12 pasas para 4 galletas.

Pistas

Otra manera de resolver problemas de multiplicación

Hacer una matriz es otra manera de resolver el problema. Usa fichas para hacer una matriz de 4 filas con 3 fichas en cada fila.

Cuenta todas las fichas.

4 grupos de 3 son 12.
$4 \times 3 = 12$

Actividad

Pida a su hijo o hija que dibuje más grupos de 3 para 5, 6, 7, 8 y 9 galletas. Después, pídale que conteste preguntas como "¿Cuántas pasas se necesitan para hacer 8 galletas? ¿Qué factores debes multiplicar para hallar la respuesta?".

© Houghton Mifflin Harcourt Publishing Company

Name _____

Multiply with 2 and 4

Write a multiplication sentence for the model.

1.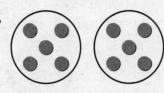

Think: There are 2 groups of 5 counters.

__2__ × __5__ = __10__

2.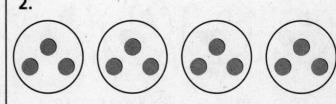

____ × ____ = ____

Find the product.

3. 2
   × 6

4. 4
   × 8

5. 2
   × 3

6. 4
   × 6

7. 4
   × 4

8. 2
   × 7

9. 4
   × 5

10. 2
    × 4

Problem Solving

11. On Monday, Steven read 9 pages of his new book. To finish the first chapter on Tuesday, he needs to read double the number of pages he read on Monday. How many pages does he need to read on Tuesday?

_____

12. Courtney's school is having a family game night. Each table has 4 players. There are 7 tables in all. How many players are at the game night?

_____

© Houghton Mifflin Harcourt Publishing Company

Lesson Check

1. Which multiplication sentence matches the model?

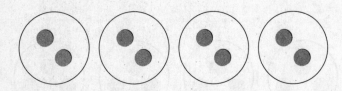

- (A) $3 \times 2 = 6$
- (B) $4 \times 2 = 8$
- (C) $4 \times 4 = 16$
- (D) $4 \times 8 = 32$

2. Find the product.

$$\begin{array}{r} 2 \\ \times\ 8 \\ \hline \end{array}$$

- (A) 10
- (B) 14
- (C) 16
- (D) 18

Spiral Review

3. Sean made a picture graph to show his friends' favorite colors. This is the key for the graph.

Each = 2 friends.

How many friends does

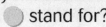 stand for? (Lesson 2.3)

- (A) 4
- (B) 8
- (C) 20
- (D) 40

4. The table shows the lengths of some walking trails.

Walking Trails	
Name	**Length (in feet)**
Mountain Trail	844
Lake Trail	792
Harmony Trail	528

How many feet longer is Mountain Trail than Harmony Trail? (Lesson 1.10)

- (A) 216 feet
- (C) 316 feet
- (B) 264 feet
- (D) 528 feet

5. Find the sum. (Lesson 1.7)

$$\begin{array}{r} 527 \\ +\ 154 \\ \hline \end{array}$$

- (A) 373
- (B) 581
- (C) 671
- (D) 681

6. A bar graph shows that sports books received 9 votes. If the scale is 0 to 20 by twos, where should the bar end for the sports books? (Lesson 2.5)

- (A) between 8 and 10
- (B) on 10
- (C) on 8
- (D) between 6 and 8

© Houghton Mifflin Harcourt Publishing Company

Name _____

Multiply with 5 and 10

Find the product.

1. $5 \times 7 =$ __35__   **2.** $5 \times 1 =$ _____   **3.** $2 \times 10 =$ _____   **4.** _____ $= 8 \times 5$

5. $1 \times 10 =$ _____   **6.** _____ $= 4 \times 5$ **7.** $5 \times 10 =$ _____   **8.** $7 \times 5 =$ _____

9. _____ $= 5 \times 5$ **10.** $5 \times 8 =$ _____   **11.** _____ $= 5 \times 9$ **12.** $10 \times 0 =$ _____

13. $\begin{array}{r} 5 \\ \times\ 6 \\ \hline \end{array}$ **14.** $\begin{array}{r} 10 \\ \times\ 7 \\ \hline \end{array}$ **15.** $\begin{array}{r} 5 \\ \times\ 3 \\ \hline \end{array}$ **16.** $\begin{array}{r} 10 \\ \times\ 4 \\ \hline \end{array}$

17. $\begin{array}{r} 5 \\ \times\ 0 \\ \hline \end{array}$ **18.** $\begin{array}{r} 10 \\ \times\ 8 \\ \hline \end{array}$ **19.** $\begin{array}{r} 5 \\ \times\ 2 \\ \hline \end{array}$ **20.** $\begin{array}{r} 10 \\ \times\ 6 \\ \hline \end{array}$

Problem Solving  REAL WORLD

21. Ginger takes 10 nickels to buy some pencils at the school store. How many cents does Ginger have to spend?

_____

22. The gym at Evergreen School has three basketball courts. There are 5 players on each of the courts. How many players are there in all?

_____

© Houghton Mifflin Harcourt Publishing Company

Lesson Check

1. Mrs. Hinely grows roses. There are 6 roses on each of her 10 rose bushes. How many roses in all are on Mrs. Hinely's rose bushes?

 (A) 16   (C) 60

 (B) 54   (D) 66

2. Find the product.

$$\begin{array}{r} 5 \\ \times\ 8 \\ \hline \end{array}$$

 (A) 8   (C) 35

 (B) 16   (D) 40

Spiral Review

3. Mr. Miller's class voted on where to go for a field trip. Use the picture graph to find which choice had the most votes. **(Lesson 2.2)**

Field Trip Choices	
Science Center	★★
Aquarium	★★★⯪
Zoo	★★★★
Museum	★★

Key: Each ★ = 2 votes.

 (A) Science Center   (C) Zoo

 (B) Aquarium   (D) Museum

4. Zack made this table for his survey.

Favorite Juice	
Flavor	Votes
Grape	16
Orange	10
Berry	9
Apple	12

How many students were surveyed in all? **(Lesson 2.6)**

 (A) 38

 (B) 43

 (C) 47

 (D) 49

5. Which of the following numbers is even? **(Lesson 1.1)**

$$25,\ 28,\ 31,\ 37$$

 (A) 25   (C) 31

 (B) 28   (D) 37

6. Estimate the sum. **(Lesson 1.3)**

$$\begin{array}{r} 479 \\ +\ \ 89 \\ \hline \end{array}$$

 (A) 300   (C) 500

 (B) 400   (D) 600

© Houghton Mifflin Harcourt Publishing Company

Name _____

Multiply with 3 and 6

Find the product.

1. $6 \times 4 = \underline{24}$ **2.** $3 \times 7 = \underline{}$ **3.** $\underline{} = 2 \times 6$ **4.** $\underline{} = 3 \times 5$

Think: You can use
doubles.
$3 \times 4 = 12$
$12 + 12 = 24$

5. $1 \times 3 = \underline{}$ **6.** $\underline{} = 6 \times 8$ **7.** $3 \times 9 = \underline{}$ **8.** $\underline{} = 6 \times 6$

9. $\begin{array}{r} 4 \\ \times\ 3 \\ \hline \end{array}$ **10.** $\begin{array}{r} 6 \\ \times\ 5 \\ \hline \end{array}$ **11.** $\begin{array}{r} 2 \\ \times\ 3 \\ \hline \end{array}$ **12.** $\begin{array}{r} 6 \\ \times\ 3 \\ \hline \end{array}$

13. $\begin{array}{r} 10 \\ \times\ 6 \\ \hline \end{array}$ **14.** $\begin{array}{r} 3 \\ \times\ 6 \\ \hline \end{array}$ **15.** $\begin{array}{r} 7 \\ \times\ 6 \\ \hline \end{array}$ **16.** $\begin{array}{r} 3 \\ \times\ 0 \\ \hline \end{array}$

17. $\begin{array}{r} 9 \\ \times\ 6 \\ \hline \end{array}$ **18.** $\begin{array}{r} 3 \\ \times\ 3 \\ \hline \end{array}$ **19.** $\begin{array}{r} 10 \\ \times\ 3 \\ \hline \end{array}$ **20.** $\begin{array}{r} 1 \\ \times\ 6 \\ \hline \end{array}$

Problem Solving  REAL WORLD

21. James got 3 hits in each of his baseball games. He has played 4 baseball games. How many hits has he had in all?

_____

22. Mrs. Burns is buying muffins. There are 6 muffins in each box. If she buys 5 boxes, how many muffins will she buy?

_____

© Houghton Mifflin Harcourt Publishing Company

Lesson Check

1. Paco buys a carton of eggs. The carton has 2 rows of eggs. There are 6 eggs in each row. How many eggs are in the carton?

   (A) 8   (C) 14

   (B) 12   (D) 24

2. Find the product.

   $$\begin{array}{r} 9 \\ \times\ 3 \\ \hline \end{array}$$

   (A) 18   (C) 27

   (B) 24   (D) 36

Spiral Review

3. Find the difference. (Lesson 1.10)

   $$\begin{array}{r} 568 \\ -\ 283 \\ \hline \end{array}$$

   (A) 285   (C) 385

   (B) 325   (D) 851

4. Dwight made double the number of baskets in the second half of the basketball game than in the first half. He made 5 baskets in the first half. How many baskets did he make in the second half? (Lesson 4.1)

   (A) 7   (C) 10

   (B) 9   (D) 20

5. In Jane's picture graph, the symbol ☺ represents two students. One row in the picture graph has 8 symbols. How many students does that represent?

   (Lesson 2.3)

   (A) 40

   (B) 32

   (C) 24

   (D) 16

6. What multiplication sentence does this array show? (Lesson 3.5)

   (A) $5 \times 6 = 30$

   (B) $6 \times 6 = 36$

   (C) $5 \times 5 = 25$

   (D) $1 \times 6 = 6$

© Houghton Mifflin Harcourt Publishing Company

Name _____

Distributive Property

Write one way to break apart the array.
Then find the product.

1.

$(3 \times 7) + (3 \times 7)$

42

2.

_____

_____

3.

_____

_____

4.

_____

_____

Problem Solving REAL WORLD

5. There are 2 rows of 8 chairs set up in the library for a puppet show. How many chairs are there in all? Use the Distributive Property to solve.

_____

6. A marching band has 4 rows of trumpeters with 10 trumpeters in each row. How many trumpeters are in the marching band? Use the Distributive Property to solve.

_____

© Houghton Mifflin Harcourt Publishing Company

Lesson Check

1. Which number sentence is an example of the Distributive Property?

   (A) $7 \times 6 = 6 \times 7$

   (B) $7 \times (2 \times 3) = (7 \times 2) \times 3$

   (C) $7 \times 6 = (7 \times 3) + (7 \times 3)$

   (D) $7 + 6 = 7 + 3 + 3$

2. What is one way to break apart the array?

   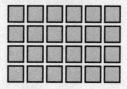

   (A) $(2 \times 6) + (2 \times 6)$

   (B) $(4 \times 2) + (4 \times 2)$

   (C) $(4 \times 4) + (4 \times 4)$

   (D) $(6 \times 3) + (6 \times 3)$

Spiral Review

3. The school auditorium has 448 chairs set out for the third-grade performance. What is 448 rounded to the nearest ten? (Lesson 1.2)

   (A) 500   (C) 450

   (B) 440   (D) 400

4. Find the difference. (Lesson 1.11)

   $$400 - 296$$

   (A) 104   (C) 204

   (B) 114   (D) 296

5. There are 622 fruit snacks in one crate and 186 in another crate. How many fruit snacks are there in all? (Lesson 1.7)

   $$622 + 186$$

   (A) 436

   (B) 708

   (C) 768

   (D) 808

6. Which sport do 6 students play? (Lesson 2.4)

   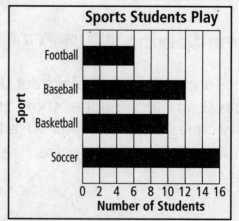

   (A) Football   (C) Basketball

   (B) Baseball   (D) Soccer

© Houghton Mifflin Harcourt Publishing Company

Name _____

Multiply with 7

Find the product.

1. $6 \times 7 = \underline{42}$ **2.** _____ $= 7 \times 9$ **3.** _____ $= 1 \times 7$ **4.** $3 \times 7 =$ _____

5. $7 \times 7 =$ _____   **6.** _____ $= 2 \times 7$ **7.** $7 \times 8 =$ _____   **8.** _____ $= 4 \times 7$

9. $\begin{array}{r} 7 \\ \times\ 5 \\ \hline \end{array}$ **10.** $\begin{array}{r} 7 \\ \times\ 1 \\ \hline \end{array}$ **11.** $\begin{array}{r} 6 \\ \times\ 7 \\ \hline \end{array}$ **12.** $\begin{array}{r} 7 \\ \times\ 4 \\ \hline \end{array}$ **13.** $\begin{array}{r} 2 \\ \times\ 7 \\ \hline \end{array}$

14. $\begin{array}{r} 10 \\ \times\ 7 \\ \hline \end{array}$ **15.** $\begin{array}{r} 3 \\ \times\ 7 \\ \hline \end{array}$ **16.** $\begin{array}{r} 7 \\ \times\ 9 \\ \hline \end{array}$ **17.** $\begin{array}{r} 8 \\ \times\ 7 \\ \hline \end{array}$ **18.** $\begin{array}{r} 7 \\ \times\ 0 \\ \hline \end{array}$

Problem Solving REAL WORLD

19. Julie buys a pair of earrings for $7. Now she would like to buy the same earrings for 2 of her friends. How much will she spend for all 3 pairs of earrings?

_____

20. Owen and his family will go camping in 8 weeks. There are 7 days in 1 week. How many days are in 8 weeks?

_____

© Houghton Mifflin Harcourt Publishing Company

Lesson Check

1. Find the product.

$$7 \times 8$$

- Ⓐ 54
- Ⓑ 56
- Ⓒ 64
- Ⓓ 66

2. What product does the array show?

- Ⓐ 14
- Ⓑ 17
- Ⓒ 21
- Ⓓ 24

Spiral Review

3. Which statement is true about the numbers below? **(Lesson 1.1)**

6, 12, 18, 24, 30

- Ⓐ All of the numbers are odd.
- Ⓑ Some of the numbers are odd.
- Ⓒ All of the numbers are even.
- Ⓓ Some of the numbers are even.

4. How many more people chose retriever than poodle? **(Lesson 2.1)**

Favorite Breed of Dog	
Dog	**Number**
Shepherd	58
Retriever	65
Poodle	26

- Ⓐ 31
- Ⓒ 41
- Ⓑ 39
- Ⓓ 49

5. What is 94 rounded to the nearest ten? **(Lesson 1.2)**

- Ⓐ 90
- Ⓑ 94
- Ⓒ 95
- Ⓓ 100

6. Jack has 5 craft sticks. He needs 4 times that number for a project. How many craft sticks does Jack need altogether? **(Lesson 4.2)**

- Ⓐ 9
- Ⓑ 16
- Ⓒ 20
- Ⓓ 24

© Houghton Mifflin Harcourt Publishing Company

Name _____

Associative Property of Multiplication

**Write another way to group the factors.
Then find the product.**

1. $(3 \times 2) \times 5$

$\underline{\quad 3 \times (2 \times 5) \quad}$

$\underline{\qquad 30 \qquad}$

2. $(4 \times 3) \times 2$

_____

_____

3. $2 \times (2 \times 8)$

_____

_____

4. $9 \times (2 \times 1)$

_____

_____

5. $2 \times (3 \times 6)$

_____

_____

6. $(4 \times 2) \times 5$

_____

_____

**Use parentheses and multiplication properties.
Then, find the product.**

7. $9 \times 1 \times 5 = $ _____

8. $3 \times 3 \times 2 = $ _____

9. $2 \times 4 \times 3 = $ _____

10. $5 \times 2 \times 3 = $ _____

11. $7 \times 1 \times 5 = $ _____

12. $8 \times 2 \times 3 = $ _____

13. $7 \times 2 \times 3 = $ _____

14. $4 \times 1 \times 3 = $ _____

15. $10 \times 2 \times 4 = $ _____

Problem Solving REAL WORLD

16. Beth and Maria are going to the county fair. Admission costs $4 per person for each day. They plan to go for 3 days. How much will the girls pay in all?

_____

17. Randy's garden has 3 rows of carrots with 3 plants in each row. Next year he plans to plant 4 times the number of rows of 3 plants. How many plants will he have next year?

_____

© Houghton Mifflin Harcourt Publishing Company

Lesson Check

1. There are 2 benches in each car of a train ride. Two people ride on each bench. If a train has 5 cars, how many people in all can be on a train?

Ⓐ 4

Ⓑ 9

Ⓒ 10

Ⓓ 20

2. Crystal has 2 CDs in each box. She has 3 boxes on each of her 6 shelves. How many CDs does Crystal have in all?

Ⓐ 6

Ⓑ 12

Ⓒ 18

Ⓓ 36

Spiral Review

3. Find the sum. (Lesson 1.7)

$$472 + 186$$

Ⓐ 658

Ⓑ 648

Ⓒ 558

Ⓓ 286

4. Trevor made a picture graph to show how many minutes each student biked last week. This is his key.

Each ⚙ = 10 minutes.

What does ⚙ ⚙ ⚙ stand for? (Lesson 2.2)

Ⓐ 2 minutes   Ⓒ 20 minutes

Ⓑ 10 minutes   Ⓓ 25 minutes

5. Madison has 142 stickers in her collection. What is 142 rounded to the nearest ten? (Lesson 1.2)

Ⓐ 40

Ⓑ 140

Ⓒ 150

Ⓓ 200

6. There are 5 pages of photos. Each page has 6 photos. How many photos are there in all? (Lesson 4.2)

Ⓐ 12

Ⓑ 20

Ⓒ 24

Ⓓ 30

© Houghton Mifflin Harcourt Publishing Company

Name _____

Patterns on the Multiplication Table

Is the product even or odd? Write *even* or *odd*.

1. $2 \times 7 = $ <u>even</u>   Think: Products with 2 as a factor are even.   2. $4 \times 6 = $ _____   3. $8 \times 3 = $ _____

4. $2 \times 3 = $ _____   5. $9 \times 9 = $ _____   6. $5 \times 7 = $ _____   7. $6 \times 3 = $ _____

Use the multiplication table. Describe a pattern you see.

8. in the column for 5

_____

_____

_____

9. in the row for 10

×	0	1	2	3	4	5	6	7	8	9	10
0	0	0	0	0	0	0	0	0	0	0	0
1	0	1	2	3	4	5	6	7	8	9	10
2	0	2	4	6	8	10	12	14	16	18	20
3	0	3	6	9	12	15	18	21	24	27	30
4	0	4	8	12	16	20	24	28	32	36	40
5	0	5	10	15	20	25	30	35	40	45	50
6	0	6	12	18	24	30	36	42	48	54	60
7	0	7	14	21	28	35	42	49	56	63	70
8	0	8	16	24	32	40	48	56	64	72	80
9	0	9	18	27	36	45	54	63	72	81	90
10	0	10	20	30	40	50	60	70	80	90	100

_____

_____

10. in the rows for 3 and 6

_____

_____

Problem Solving REAL WORLD

11. Carl shades a row in the multiplication table. The products in the row are all even. The ones digits in the products repeat 0, 4, 8, 2, 6. What row does Carl shade?

_____

12. Jenna says that no row or column contains products with only odd numbers. Do you agree? **Explain.**

_____

_____

_____

© Houghton Mifflin Harcourt Publishing Company

Lesson Check

1. Which has an even product?

   Ⓐ 1 × 9
   Ⓑ 3 × 3
   Ⓒ 5 × 7
   Ⓓ 4 × 9

2. Which describes this pattern?

   10, 15, 20, 25, 30

   Ⓐ Even and then odd
   Ⓑ Add 10.
   Ⓒ Subtract 5.
   Ⓓ Multiply by 5.

Spiral Review

3. Lexi has 2 cans of tennis balls. There are 3 tennis balls in each can. She buys 2 more cans. How many tennis balls does she now have in all? (Lesson 4.6)

   Ⓐ 12
   Ⓑ 9
   Ⓒ 7
   Ⓓ 6

4. Use the picture graph.

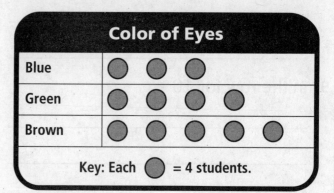

   How many students have green eyes? (Lesson 2.2)

   Ⓐ 4   Ⓒ 12
   Ⓑ 8   Ⓓ 16

5. Sasha bought 3 boxes of pencils. If each box has 6 pencils, how many pencils did Sasha buy in all?
   (Lesson 4.3)

   Ⓐ 9   Ⓒ 18
   Ⓑ 12   Ⓓ 24

6. Find the sum. (Lesson 1.7)

   219
   + 763

   Ⓐ 992   Ⓒ 976
   Ⓑ 982   Ⓓ 972

© Houghton Mifflin Harcourt Publishing Company

Name _____

Multiply with 8

Find the product.

1. $8 \times 10 =$ __80__   2. $8 \times 8 =$ ____   3. $8 \times 5 =$ ____   4. $3 \times 8 =$ ____

5. ____ $= 4 \times 8$ 6. $8 \times 7 =$ ____   7. $6 \times 8 =$ ____   8. ____ $= 9 \times 8$

9. $\begin{array}{r} 8 \\ \times\ 2 \\ \hline \end{array}$ 10. $\begin{array}{r} 6 \\ \times\ 8 \\ \hline \end{array}$ 11. $\begin{array}{r} 8 \\ \times\ 7 \\ \hline \end{array}$ 12. $\begin{array}{r} 0 \\ \times\ 8 \\ \hline \end{array}$ 13. $\begin{array}{r} 8 \\ \times\ 5 \\ \hline \end{array}$

14. $\begin{array}{r} 8 \\ \times\ 8 \\ \hline \end{array}$ 15. $\begin{array}{r} 9 \\ \times\ 8 \\ \hline \end{array}$ 16. $\begin{array}{r} 8 \\ \times\ 3 \\ \hline \end{array}$ 17. $\begin{array}{r} 8 \\ \times\ 1 \\ \hline \end{array}$ 18. $\begin{array}{r} 4 \\ \times\ 8 \\ \hline \end{array}$

Problem Solving REAL WORLD

19. There are 6 teams in the basketball league. Each team has 8 players. How many players are there in all?

_____

20. Lynn has 4 stacks of quarters. There are 8 quarters in each stack. How many quarters does Lynn have in all?

_____

21. Tomas is packing 7 baskets for a fair. He is placing 8 apples in each basket. How many apples are there in all?

_____

22. There are 10 pencils in each box. If Jenna buys 8 boxes, how many pencils will she buy?

_____

© Houghton Mifflin Harcourt Publishing Company

Lesson Check

1. Find the product.

$5 \times 8 =$ ■

Ⓐ 30

Ⓑ 32

Ⓒ 42

Ⓓ 40

2. There are 7 tarantulas in the spider exhibit at the zoo. Each tarantula has 8 legs. How many legs do the 7 tarantulas have in all?

Ⓐ 15   Ⓒ 56

Ⓑ 49   Ⓓ 63

Spiral Review

3. Find the difference. (Lesson 1.9)

$$652$$
$$-\ \ 99$$

Ⓐ 99

Ⓑ 552

Ⓒ 553

Ⓓ 653

4. The school library received an order of 232 new books. What is 232 rounded to the nearest ten? (Lesson 1.8)

Ⓐ 200

Ⓑ 230

Ⓒ 240

Ⓓ 300

5. Sam's picture graph shows that 8 students chose pizza as their favorite lunch. This is the key for the graph.

Each ☺ = 2 students.
How many ☺ should be next to pizza on Sam's graph? (Lesson 2.2)

Ⓐ 2

Ⓑ 4

Ⓒ 6

Ⓓ 8

6. Tashia buys 5 packages of oranges. Each package has 4 oranges. How many oranges in all does Tashia buy? (Lesson 4.2)

Ⓐ 1

Ⓑ 9

Ⓒ 20

Ⓓ 25

© Houghton Mifflin Harcourt Publishing Company

Name _____

Multiply with 9

Find the product.

1. $10 \times 9 =$ __90__   **2.** $2 \times 9 =$ ____   **3.** $9 \times 4 =$ ____   **4.** $0 \times 9 =$ ____

5. $1 \times 9 =$ ____   **6.** $8 \times 9 =$ ____   **7.** $9 \times 5 =$ ____   **8.** $6 \times 9 =$ ____

9. $\begin{array}{r} 9 \\ \times\ 4 \\ \hline \end{array}$ **10.** $\begin{array}{r} 5 \\ \times\ 9 \\ \hline \end{array}$ **11.** $\begin{array}{r} 9 \\ \times\ 7 \\ \hline \end{array}$ **12.** $\begin{array}{r} 2 \\ \times\ 9 \\ \hline \end{array}$ **13.** $\begin{array}{r} 9 \\ \times\ 9 \\ \hline \end{array}$

14. $\begin{array}{r} 10 \\ \times\ 9 \\ \hline \end{array}$ **15.** $\begin{array}{r} 3 \\ \times\ 9 \\ \hline \end{array}$ **16.** $\begin{array}{r} 9 \\ \times\ 8 \\ \hline \end{array}$ **17.** $\begin{array}{r} 6 \\ \times\ 9 \\ \hline \end{array}$ **18.** $\begin{array}{r} 9 \\ \times\ 1 \\ \hline \end{array}$

Problem Solving  REAL WORLD

19. There are 9 positions on the softball team. Three people are trying out for each position. How many people in all are trying out?

_____

20. Carlos bought a book for $9. Now he would like to buy 4 other books for the same price. How much will he have to pay in all for the other 4 books?

_____

© Houghton Mifflin Harcourt Publishing Company

Lesson Check

1. Find the product.

$7 \times 9 = \blacksquare$

Ⓐ 63

Ⓑ 56

Ⓒ 45

Ⓓ 36

2. Clare buys 5 tickets for the high school musical. Each ticket costs $9. How much do the tickets cost in all?

Ⓐ $36   Ⓒ $45

Ⓑ $40   Ⓓ $52

Spiral Review

3. The table shows the hair color of girls in Kim's class. How many girls have brown hair? (Lesson 2.1)

Kim's Class	
Hair Color	Number of Girls
Brown	⅃⅃⅃⅃ I
Black	III
Blonde	IIII
Red	I

Ⓐ 1   Ⓒ 4

Ⓑ 3   Ⓓ 6

4. Miles picked up 9 shirts from the dry cleaners. It costs $4 to clean each shirt. How much did Miles spend to have all the shirts cleaned? (Lesson 4.1)

Ⓐ $13

Ⓑ $22

Ⓒ $36

Ⓓ $45

5. In a picture graph, each picture of a baseball is equal to 5 games won by a team. The row for the Falcons has 7 baseballs. How many games have the Falcons won? (Lesson 2.2)

Ⓐ 40   Ⓒ 12

Ⓑ 35   Ⓓ 7

6. An array has 8 rows with 4 circles in each row. How many circles are in the array? (Lesson 4.8)

Ⓐ 12   Ⓒ 32

Ⓑ 24   Ⓓ 36

© Houghton Mifflin Harcourt Publishing Company

Name _____

Problem Solving • Multiplication

Solve.

1. Henry has a new album for his baseball cards. He uses pages that hold 6 cards and pages that hold 3 cards. If Henry has 36 cards, how many different ways can he put them in his album?

Pages with 6 Cards	1	2	3	4	5
Pages with 3 Cards	10	8	6	4	2
Total Cards	36	36	36	36	36

Henry can put the cards in his album __5__ ways.

---

2. Ms. Hernandez has 17 tomato plants that she wants to plant in rows. She will put 2 plants in some rows and 1 plant in the others. How many different ways can she plant the tomato plants? Make a table to solve.

Rows with 2 Plants	
Rows with 1 Plant	
Total Plants	

Ms. Hernandez can plant the tomato plants _____ ways.

---

3. Bianca has a total of 25¢. She has some nickels and pennies. How many different combinations of nickels and pennies could Bianca have? Make a table to solve.

Number of Nickels	
Number of Pennies	
Total Value	

Bianca could have _____ combinations of 25¢.

© Houghton Mifflin Harcourt Publishing Company

Lesson Check

1. The table shows different ways that Cameron can display his 12 model cars on shelves. How many shelves will display 2 cars if 8 of the shelves each display 1 car?

Shelves with 1 Car	2	4	6	8	10
Shelves with 2 Cars	5	4	3	■	■
Total cars	12	12	12	12	12

Ⓐ 1   Ⓒ 3

Ⓑ 2   Ⓓ 4

Spiral Review

2. Find the sum. (Lesson 1.6)

$$317 + 151$$

Ⓐ 166   Ⓒ 468

Ⓑ 268   Ⓓ 568

3. The school cafeteria has an order for 238 hot lunches. What is 238 rounded to the nearest ten? (Lesson 1.2)

Ⓐ 300   Ⓒ 230

Ⓑ 240   Ⓓ 200

4. Tyler made a picture graph to show students' favorite colors. This is the key for his graph.

     Each ● = 3 votes.

If 12 students voted for green, how many ● should there be in the green row of the graph? (Lesson 2.2)

Ⓐ 3   Ⓒ 9

Ⓑ 4   Ⓓ 12

5. There are 5 bikes in each bike rack at the school. There are 6 bike racks. How many bikes in all are in the bike racks? (Lesson 4.2)

Ⓐ 11

Ⓑ 24

Ⓒ 25

Ⓓ 30

© Houghton Mifflin Harcourt Publishing Company

Name _____

Chapter 4 Extra Practice

Lessons 4.1 - 4.2

Find the product.

1. $4 \times 2 =$ _____   **2.** $8 \times 5 =$ _____   **3.** $10 \times 7 =$ _____   **4.** $2 \times 9 =$ _____

5.   6   **6.**   5   **7.**   2   **8.**   4
     $\times\ 10$ $\times\ 7$ $\times\ 10$ $\times\ 5$
     _____   _____   _____   _____

Lessons 4.3 - 4.5

Find the product.

1.   6   **2.**   3   **3.**   7   **4.**   8
     $\times\ 2$ $\times\ 9$ $\times\ 3$ $\times\ 6$
     _____   _____   _____   _____

Write one way to break apart the array. Then find the product.

5.

_____

_____

Find the product.

6. $5 \times 7 =$ _____   **7.** $2 \times 6 =$ _____   **8.** $4 \times 7 =$ _____   **9.** $8 \times 3 =$ _____

10. Abby has 5 stacks of cards with 7 cards in each stack. How many cards does she have in all?

_____

11. Noah has 3 sisters. He gave 6 balloons to each sister. How many balloons did Noah give away in all?

_____

© Houghton Mifflin Harcourt Publishing Company

Lesson 4.6

Write another way to group the factors. Then find the product.

1. $(3 \times 2) \times 4$

2. $2 \times (5 \times 3)$

3. $(1 \times 4) \times 2$

_____

_____

_____

_____

_____

_____

Lesson 4.7

Is the product even or odd?
Write *even* or *odd*.

1. $6 \times 6 =$ _____

2. $2 \times 3 =$ _____

3. $3 \times 9 =$ _____

×	0	1	2	3	4	5	6	7	8	9	10
0	0	0	0	0	0	0	0	0	0	0	0
1	0	1	2	3	4	5	6	7	8	9	10
2	0	2	4	6	8	10	12	14	16	18	20
3	0	3	6	9	12	15	18	21	24	27	30
4	0	4	8	12	16	20	24	28	32	36	40
5	0	5	10	15	20	25	30	35	40	45	50
6	0	6	12	18	24	30	36	42	48	54	60

Lessons 4.8 – 4.9

Find the product.

1. $8 \times 2 =$ _____

2. $5 \times 9 =$ _____

3. _____ $= 3 \times 9$

4. $4 \times 8 =$ _____

5. _____ $= 9 \times 4$

6. $6 \times 8 =$ _____

7. $9 \times 7 =$ _____

8. _____ $= 8 \times 7$

Lesson 4.10

1. Leo has a total of 45¢. He has some dimes and pennies. How many different combinations of dimes and pennies could Leo have? Make a table to solve.

Leo could have _____ combinations of 45¢.

Number of Dimes				
Number of Pennies				
Total Value				

© Houghton Mifflin Harcourt Publishing Company

Chapter 5

School-Home Letter

Vocabulary

equation A number sentence that uses the equal sign to show that two amounts are equal

factor A number that is multiplied by another number to find a product

multiple A number that is the product of two counting numbers

product The answer to a multiplication problem

Dear Family,

During the next few weeks, our math class will be learning more about multiplication. We will learn strategies for finding an unknown factor and for multiplying with multiples of 10.

You can expect to see homework that provides practice with strategies for multiplying with multiples of 10.

Here is a sample of how your child will be taught to use a number line to multiply.

🔑 MODEL Use a number line to find 3 × 50.

Think: 50 = 5 tens

STEP 1

Draw a number line and write the labels for multiples of 10.

STEP 2

Draw jumps on the number line to show 3 groups of 5 tens.

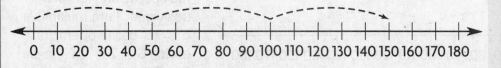

So, 3 × 50 = 150.

Tips

Using Place Value to Multiply

Using a multiplication fact and place value is another way to multiply by a multiple of 10. To multiply 6 × 70, use the basic fact 6 × 7 = 42. Think: 6 × 7 tens = 42 tens, or 420.

Activity

Help your child draw and use models to multiply with multiples of 10. Ask your child to solve problems such as, "There are 6 chocolate chips on one cookie. How do you multiply to find the number of chocolate chips on 20 cookies?"

© Houghton Mifflin Harcourt Publishing Company

Carta para la casa

Vocabulario

ecuación Una expresión numérica que muestra que dos cantidades son iguales

factor Un número que se multiplica por otro número para hallar un producto

múltiplo Un número que es el producto de dos números naturales distintos de cero

producto El resultado en un problema de multiplicación

Querida familia,

Durante las próximas semanas, en la clase de matemáticas aprenderemos más sobre la multiplicación. Aprenderemos estrategias para hallar un factor desconocido y para multiplicar por múltiplos de 10.

Llevaré a casa tareas para practicar estrategias para multiplicar con múltiplos de 10.

Este es un ejemplo de cómo usaremos una recta numérica para multiplicar.

🔒 MODELO Usar una recta numérica para hallar 3 × 50

Piensa: 50 = 5 decenas

PASO 1

Traza una recta numérica y escribe los rótulos para los múltiplos de 10.

PASO 2

Dibuja saltos en la recta numérica para mostrar 3 grupos de 5 decenas.

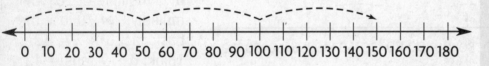

Por tanto, 3 × 50 = 150.

Pistas

Usar el valor posicional para multiplicar

Otra forma de multiplicar por un múltiplo de 10 es usar una operación de multiplicación y el valor posicional. Para multiplicar 6 × 70, usa la operación básica 6 × 7 = 42. Piensa: 6 × 7 decenas = 42 decenas, o 420.

Actividad

Ayude a su hijo/a a dibujar y usar modelos para multiplicar con múltiplos de 10. Pida a su hijo/a que resuelva problemas como "Hay 6 chispas de chocolate en una galleta. ¿Cómo multiplicas para hallar el número de chispas de chocolate que hay en 20 galletas?"

© Houghton Mifflin Harcourt Publishing Company

Name _____

Describe Patterns

Describe a pattern for the table. Then complete the table.

1.

Pans	1	2	3	4	5
Muffins	6	12	18	**24**	**30**

Add 6 muffins for each pan; Multiply the number of pans by 6.

2.

Wagons	2	3	4	5	6
Wheels	8	12	16		

_____

_____

_____

3.

Vases	Flowers
2	14
3	
4	28
5	
6	42

_____

_____

4.

Spiders	Legs
1	8
2	
3	24
4	
5	40

_____

_____

Problem Solving REAL WORLD

5. Caleb buys 5 cartons of yogurt. Each carton has 8 yogurt cups. How many yogurt cups does Caleb buy?

_____

6. Libby bought 4 packages of pencils. Each package has 6 pencils. How many pencils did Libby buy?

_____

© Houghton Mifflin Harcourt Publishing Company

Lesson Check

1. Which of the following describes a pattern in the table?

Tables	1	2	3	4	5
Chairs	5	10	15	20	25

- (A) Multiply by 3.
- (C) Add 1.
- (B) Multiply by 5.
- (D) Add 4.

2. Which number completes this table?

Butterflies	3	4	5	6	7
Wings	12	16	20	▪	28

- (A) 30
- (C) 24
- (B) 26
- (D) 22

Spiral Review

3. Jennilee buys 7 packs of crayons. There are 6 crayons in each pack. How many crayons does Jennilee buy in all? (Lesson 4.3)

- (A) 13
- (B) 36
- (C) 42
- (D) 48

4. Maverick has 5 books of circus tickets. Each book has 5 tickets. How many tickets does Maverick have in all? (Lesson 4.2)

- (A) 10
- (B) 15
- (C) 20
- (D) 25

5. Bailey walked his dog 2 times each day for 9 days. How many times did Bailey walk his dog in all? (Lesson 4.9)

- (A) 9
- (B) 11
- (C) 18
- (D) 27

6. Drew's Tree Company delivers pear trees in groups of 4. Yesterday, the company delivered 8 groups of pear trees. How many pear trees were delivered in all? (Lesson 4.8)

- (A) 12
- (B) 16
- (C) 24
- (D) 32

© Houghton Mifflin Harcourt Publishing Company

Name _____

Find Unknown Factors

Find the unknown factor.

1. $n \times 3 = 12$

Think: How many groups of 3 equal 12?

$n = \underline{\;4\;}$

2. $s \times 8 = 64$

$s = \underline{\quad}$

3. $21 = 7 \times n$

$n = \underline{\quad}$

4. $y \times 2 = 18$

$y = \underline{\quad}$

5. $5 \times p = 10$

$p = \underline{\quad}$

6. $56 = 8 \times t$

$t = \underline{\quad}$

7. $m \times 4 = 28$

$m = \underline{\quad}$

8. $\star \times 1 = 9$

$\star = \underline{\quad}$

9. $18 = 6 \times r$

$r = \underline{\quad}$

10. $u \times 5 = 30$

$u = \underline{\quad}$

11. $4 \times \blacksquare = 24$

$\blacksquare = \underline{\quad}$

12. $w \times 7 = 35$

$w = \underline{\quad}$

13. $b \times 6 = 54$

$b = \underline{\quad}$

14. $5 \times \blacktriangle = 40$

$\blacktriangle = \underline{\quad}$

15. $30 = d \times 3$

$d = \underline{\quad}$

16. $7 \times k = 42$

$k = \underline{\quad}$

Problem Solving REAL WORLD

17. Carmen spent $42 for 6 hats. How much did each hat cost?

_____

18. Mark has a baking tray with 24 cupcakes. The cupcakes are arranged in 4 equal rows. How many cupcakes are in each row?

_____

© Houghton Mifflin Harcourt Publishing Company

Lesson Check

1. What is the unknown factor?

   $b \times 7 = 56$

   (A) 6

   (B) 7

   (C) 8

   (D) 9

2. What is the unknown factor shown by this array?

   $3 \times \blacksquare = 24$

   (A) 3   (C) 8

   (B) 6   (D) 9

Spiral Review

3. Which is an example of the Commutative Property of Multiplication? (Lesson 3.6)

   (A) $6 + 4 = 4 + 6$

   (B) $4 \times 6 = 6 \times 4$

   (C) $4 \times 3 = 4 + 8$

   (D) $3 \times 6 = 9 \times 2$

4. Find the product. (Lesson 4.6)

   $5 \times (4 \times 2)$

   (A) 13

   (B) 22

   (C) 40

   (D) 80

5. Which number sentence is an example of the Distributive Property? (Lesson 4.4)

   (A) $4 \times 7 = (4 \times 3) + (4 \times 4)$

   (B) $4 \times 7 = 7 \times 4$

   (C) $4 \times 7 = 28$

   (D) $7 \times 4 = 15 + 13$

6. In a group of 10 boys, each boy had 2 hats. How many hats did they have in all? (Lesson 4.2)

   (A) 5

   (B) 12

   (C) 20

   (D) 40

© Houghton Mifflin Harcourt Publishing Company

Name _____

Problem Solving • Use the Distributive Property

Read each problem and solve.

1. Each time a student turns in a perfect spelling test, Ms. Ricks puts an achievement square on the bulletin board. There are 6 rows of squares on the bulletin board. Each row has 30 squares. How many perfect spelling tests have been turned in?

   Think: $6 \times 30 = 6 \times (10 + 10 + 10)$

   $= 60 + 60 + 60 = 180$

   180 spelling tests

2. Norma practices violin for 50 minutes every day. How many minutes does Norma practice violin in 7 days?

   _____

3. A kitchen designer is creating a new backsplash for the wall behind a kitchen sink. The backsplash will have 5 rows of tiles. Each row will have 20 tiles. How many tiles are needed for the entire backsplash?

   _____

4. A bowling alley keeps shoes in rows of cubbyholes. There are 9 rows of cubbyholes, with 20 cubbyholes in each row. If there is a pair of shoes in every cubbyhole, how many pairs of shoes are there?

   _____

5. The third-grade students are traveling to the science museum in 8 buses. There are 40 students on each bus. How many students are going to the museum?

   _____

© Houghton Mifflin Harcourt Publishing Company

Lesson Check

1. Each snack pack holds 20 crackers. How many crackers in all are there in 4 snack packs?
   - (A) 60
   - (B) 80
   - (C) 100
   - (D) 800

2. A machine makes 70 springs each hour. How many springs will the machine make in 8 hours?
   - (A) 500
   - (B) 520
   - (C) 540
   - (D) 560

Spiral Review

3. Lila read 142 pages on Friday and 168 pages on Saturday. Which is the best estimate of how many pages Lila read on Friday and Saturday combined? (Lesson 1.3)
   - (A) 100
   - (C) 300
   - (B) 200
   - (D) 400

4. Jessica wrote $6 + 6 + 6 + 6$ on the board. Which is another way to show $6 + 6 + 6 + 6$? (Lesson 3.2)
   - (A) 4×4
   - (C) $4 \times 4 \times 6$
   - (B) 4×6
   - (D) 6×6

Use the line plot for 5–6.

5. Eliot made a line plot to record the number of birds he saw at his bird feeder. How many more sparrows than blue jays did he see? (Lesson 2.7)
   - (A) 2
   - (C) 4
   - (B) 3
   - (D) 5

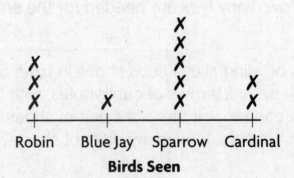

Birds Seen

6. How many robins and cardinals combined did Eliot see? (Lesson 2.7)
   - (A) 2
   - (C) 4
   - (B) 3
   - (D) 5

© Houghton Mifflin Harcourt Publishing Company

Name _____

Multiplication Strategies with Multiples of 10

Use a number line to find the product.

1. $2 \times 40 =$ __**80**__

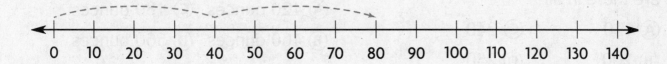

2. $4 \times 30 =$ _____

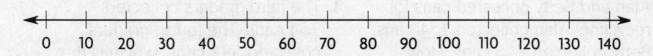

Use place value to find the product.

3. $5 \times 70 = 5 \times$ _____ tens

      $=$ _____ tens $=$ _____

4. $60 \times 4 =$ _____ tens $\times 4$

      $=$ _____ tens $=$ _____

5. $7 \times 30 = 7 \times$ _____ tens

      $=$ _____ tens $=$ _____

6. $90 \times 3 =$ _____ tens $\times 3$

      $=$ _____ tens $=$ _____

Problem Solving REAL WORLD

7. One exhibit at the aquarium has 5 fish tanks. Each fish tank holds 50 gallons of water. How much water do the 5 tanks hold in all?

_____

8. In another aquarium display, there are 40 fish in each of 7 large tanks. How many fish are in the display in all?

_____

© Houghton Mifflin Harcourt Publishing Company

Lesson Check

1. Each bag of pattern blocks contains 50 blocks. To make a class pattern, the teacher combines 4 bags of blocks. How many pattern blocks are there in all?

(A) 20

(C) 240

(B) 200

(D) 250

2. A deli received 8 blocks of cheese. Each block of cheese weighs 60 ounces. What is the total weight of the cheeses?

(A) 420 ounces (C) 480 ounces

(B) 460 ounces (D) 560 ounces

Spiral Review

3. Alan and Betty collected cans for recycling. Alan collected 154 cans. Betty collected 215 cans. How many cans did they collect in all?

(Lesson 1.6)

(A) 369

(C) 469

(B) 379

(D) 479

4. The third graders collected 754 cans. The fourth graders collected 592 cans. Which is the best estimate of how many more cans the third graders collected?

(Lesson 1.8)

(A) 50

(C) 200

(B) 100

(D) 300

Use the bar graph for 5–6.

5. How many more books did Ed read than Bob? (Lesson 2.4)

(A) 2

(C) 4

(B) 3

(D) 5

6. How many books in all did the four students read in June? (Lesson 2.4)

(A) 22

(C) 26

(B) 24

(D) 36

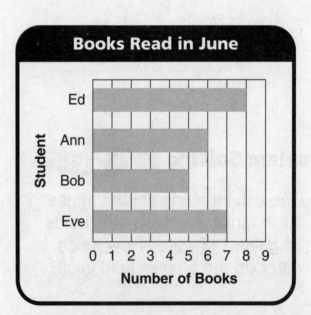

Books Read in June

© Houghton Mifflin Harcourt Publishing Company

School-Home Letter

© Houghton Mifflin Harcourt Publishing Company

Vocabulary

dividend The number that is to be divided in a division problem

dividend, divisor, quotient The parts of a division problem. There are two ways to record division.

$$10 \div 2 = 5$$

↑ dividend   ↑ divisor   ↑ quotient

divisor → $2\overline{)10}$ ← quotient

dividend

Dear Family,

During the next few weeks, our math class will be learning about division. We will learn how division is related to subtraction, and how multiplication and division are inverse operations.

You can expect to see homework that provides practice with division.

Here is a sample of how your child will be taught to use repeated subtraction to solve division problems.

🔑 MODEL Use Repeated Subtraction to Divide

This is how we will be using repeated subtraction to divide.

STEP 1

Start with the dividend and subtract the divisor until you reach 0.

$15 \div 5 = $ ___

$$\begin{array}{ccc} 15 & 10 & \\ -5 & -5 & -5 \\ \hline 10 & 5 & 0 \end{array}$$

STEP 2

Count the number of times you subtract 5.

$$\begin{array}{ccc} 15 & 10 & 5 \\ -5 & -5 & -5 \\ \hline 10 & 5 & 0 \end{array} \text{ (3 times)}$$

There are 3 groups of 5 in 15.

STEP 3

Record the quotient.

$15 \div 5 = 3$, or

$5\overline{)15}$ with quotient 3

Fifteen divided by 5 equals 3.

Tips

Counting Back on a Number Line

Counting back on a number line is another way to find a quotient. On a 0–15 number line, for example, start at 15 and count back by 5s to 0. Then count the number of jumps on a number line (3 jumps) to find that $15 \div 5 = 3$.

Activity

Display a number of objects that are divisible by 5. Have your child use repeated subtraction to solve division problems. For example: "Here are 20 crayons. I want to subtract 5 crayons at a time until there are no crayons left. How many times can I subtract?" Check answers by arranging the objects.

Capítulo

6

Carta
para la casa

Estimada familia,

Durante las próximas semanas, nuestra clase de matemáticas aprenderá sobre la división. Aprenderemos sobre cómo la división se relaciona con la resta, y cómo la multiplicación y la división son operaciones inversas.

Pueden esperar ver tareas que sirven para practicar la división.

Esta es una muestra de cómo su hijo o hija aprenderá a usar la resta repetida para resolver problemas de división.

Vocabulario

dividendo El número que se divide en un problema de división.

dividendo, divisor, cociente Las partes de un problema de división. Hay dos maneras de anotar la división.

$$10 \div 2 = 5$$

dividendo   divisor   cociente

$$5 \leftarrow \text{cociente}$$
$$\text{divisor} \rightarrow 2\overline{)10}$$
$$\uparrow$$
dividendo

🔑 MODELO Usar la resta repetida para dividir

Así es como usaremos la resta repetida para dividir.

PASO 1

Comience con el dividendo y réstele el divisor hasta llegar a 0.

$$15 \div 5 = \underline{\quad}$$

$$\begin{array}{ccc} 15 & 10 & 5 \\ \underline{-5} & \underline{-5} & \underline{-5} \\ 10 & 5 & 0 \end{array}$$

PASO 2

Cuente la cantidad de veces que restó 5.

$$\begin{array}{ccc} 15 & 10 & 5 \\ \underline{-5} & \underline{-5} & \underline{-5} \text{ (3 veces)} \\ 10 & 5 & 0 \end{array}$$

Hay 3 grupos de 5 en 15.

PASO 3

Anote el cociente.

$$15 \div 5 = 3, \text{ o}$$

$$5\overline{)15}^{\,3}$$

Quince dividido entre 5 es igual a 3.

Pistas

Contar hacia atrás en una recta numérica

Contar hacia atrás en una recta numérica es otra manera de hallar un cociente. En una recta numérica de 0–15, por ejemplo, comience en 15 y cuente hacia atrás de 5 en 5 hasta 0. Después cuente la cantidad de saltos que da en la recta numérica (3 saltos), para hallar que $15 \div 5 = 3$.

Actividad

Muestre una cantidad de objetos que sea divisible entre 5. Pida a su hijo o hija que use la resta repetida para resolver problemas de división. Por ejemplo: "Aquí hay 20 crayolas. Quiero restar 5 crayolas a la vez hasta que no queden crayolas. ¿Cuántas veces puedo restar?". Compruebe las respuestas ordenando los objetos.

© Houghton Mifflin Harcourt Publishing Company

Name _____

Problem Solving • Model Division

Solve each problem.

1. Six customers at a toy store bought 18 jump ropes. Each customer bought the same number of jump ropes. How many jump ropes did each customer buy?

   3 jump ropes

2. Hiro has 36 pictures of his summer trip. He wants to put them in an album. Each page of the album holds 4 pictures. How many pages will Hiro need for his pictures?

   _____

3. Katia has 42 crayons in a box. She buys a storage bin that has 6 sections. She puts the same number of crayons in each section. How many crayons does Katia put in each section of the storage bin?

   _____

4. Ms. Taylor's students give cards to each of the 3 class parent helpers. There are 24 cards. How many cards will each helper get if the students give an equal number of cards to each helper?

   _____

5. Jamie divides 20 baseball stickers equally among 5 of his friends. How many stickers does each friend get?

   _____

© Houghton Mifflin Harcourt Publishing Company

Lesson Check

1. Maria buys 15 apples at the store and places them into bags. She puts 5 apples into each bag. How many bags does Maria use for all the apples?

   (A) 2   (C) 4

   (B) 3   (D) 10

2. Tom's neighbor is fixing a section of his walkway. He has 32 bricks that he is placing in 8 equal rows. How many bricks will Tom's neighbor place in each row?

   (A) 3   (C) 5

   (B) 4   (D) 6

Spiral Review

3. Find the unknown factor. (Lesson 5.2)

   $$7 \times \blacksquare = 56$$

   (A) 6

   (B) 7

   (C) 8

   (D) 9

4. How many students practiced the piano more than 3 hours a week?

   (Lesson 2.7)

   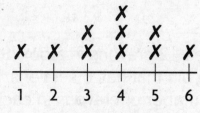

   Piano Practice Hours

   (A) 2   (C) 8

   (B) 6   (D) 10

5. Count equal groups to find how many there are. (Lesson 3.1)

   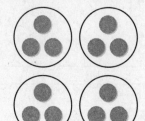

   (A) 3   (C) 12

   (B) 4   (D) 16

6. Which is another way to group the factors? (Lesson 4.6)

   $$(3 \times 2) \times 5$$

   (A) $(3 + 2) + 5$

   (B) $(3 \times 2) + 5$

   (C) $3 \times (2 + 5)$

   (D) $3 \times (2 \times 5)$

© Houghton Mifflin Harcourt Publishing Company

Name _____

Size of Equal Groups

Use counters or draw a quick picture. Make equal groups. Complete the table.

	Counters	Number of Equal Groups	Number in Each Group
1.	15	3	5
2.	21	7	
3.	28	7	
4.	32	4	
5.	9	3	
6.	18	3	
7.	20	5	
8.	16	8	
9.	35	5	
10.	24	3	

Problem Solving  REAL WORLD

11. Alicia has 12 eggs that she will use to make 4 different cookie recipes. If each recipe calls for the same number of eggs, how many eggs will she use in each recipe?

_____

12. Brett picked 27 flowers from the garden. He plans to give an equal number of flowers to each of 3 people. How many flowers will each person get?

_____

© Houghton Mifflin Harcourt Publishing Company

Lesson Check

1. Ryan has 21 pencils. He wants to put the same number of pencils in each of 3 pencil holders. How many pencils will he put in each pencil holder?

   Ⓐ 6
   Ⓑ 7
   Ⓒ 8
   Ⓓ 9

2. Corrine is setting out 24 plates on 6 tables for a dinner. She sets the same number of plates on each table. How many plates does Corrine set on each table?

   Ⓐ 3
   Ⓑ 4
   Ⓒ 5
   Ⓓ 6

Spiral Review

3. Each table has 4 legs. How many legs do 4 tables have? (Lesson 3.1)

   Ⓐ 1
   Ⓑ 8
   Ⓒ 16
   Ⓓ 20

4. Tina has 3 stacks of 5 CDs on each of 3 shelves. How many CDs does she have in all? (Lesson 4.6)

   Ⓐ 14
   Ⓑ 30
   Ⓒ 35
   Ⓓ 45

5. What is the unknown factor?
   (Lesson 5.2)

   $$7 \times \blacksquare = 35$$

   Ⓐ 4
   Ⓑ 5
   Ⓒ 6
   Ⓓ 7

6. Which of the following describes a pattern in the table? (Lesson 5.1)

Number of packs	1	2	3	4	5
Number of yo-yos	3	6	9	12	?

   Ⓐ Add 2.
   Ⓑ Multiply by 2.
   Ⓒ Multiply by 3.
   Ⓓ Add 12.

© Houghton Mifflin Harcourt Publishing Company

Name _____

Number of Equal Groups

Draw counters on your MathBoard. Then circle equal groups. Complete the table.

	Counters	Number of Equal Groups	Number in Each Group
1.	24	3	8
2.	35		7
3.	30		5
4.	16		4
5.	12		6
6.	36		9
7.	18		3
8.	15		5
9.	28		4
10.	27		3

Problem Solving REAL WORLD

11. In his bookstore, Toby places 21 books on shelves, with 7 books on each shelf. How many shelves does Toby need?

12. Mr. Holden has 32 quarters in stacks of 4 on his desk. How many stacks of quarters are on his desk?

_____ _____

© Houghton Mifflin Harcourt Publishing Company

Lesson Check

1. Ramon works at a clothing store. He puts 24 pairs of jeans into stacks of 8. How many stacks does Ramon make?

   (A) 5

   (B) 4

   (C) 3

   (D) 2

2. There are 36 people waiting in line for a hay ride. Only 6 people can ride on each wagon. If each wagon is full, how many wagons are needed for all 36 people?

   (A) 5

   (B) 6

   (C) 7

   (D) 8

Spiral Review

3. Which multiplication sentence does the array show? **(Lesson 3.5)**

   (A) 4 × 5 = 20   (C) 4 × 7 = 28

   (B) 4 × 6 = 24   (D) 4 × 8 = 32

4. Austin buys 4 boxes of nails for his project. There are 30 nails in each box. How many nails does Austin buy in all? **(Lesson 5.4)**

   (A) 12

   (B) 34

   (C) 70

   (D) 120

5. Which describes the number sentence? **(Lesson 1.1)**

   $$8 + 0 = 8$$

   (A) odd + odd = odd

   (B) Identity Property of Addition

   (C) even + even = even

   (D) Commutative Property of Addition

6. Each month for 6 months, Kelsey completes 5 paintings. How many more paintings does she need to complete before she has completed 38 paintings?

   (Lesson 4.10)

   (A) 2   (C) 8

   (B) 6   (D) 9

© Houghton Mifflin Harcourt Publishing Company

Name _____

Model with Bar Models

Write a division equation for the picture.

1.
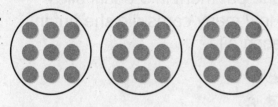

$$27 \div 3 = 9 \text{ or } 27 \div 9 = 3$$

2.

_____

3.
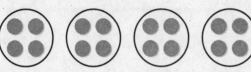

_____

4.
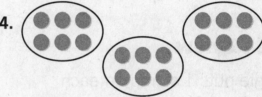

_____

Complete the bar model to solve. Then write a division equation for the bar model.

5. There are 15 postcards in 3 equal stacks. How many postcards are in each stack?

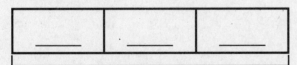

15 postcards

_____

6. There are 21 key rings. How many groups of 3 key rings can you make?

▢ groups

| 3 | | 3 |

21 key rings

_____

Problem Solving REAL WORLD

7. Jalyn collected 24 stones. She put them in 4 equal piles. How many stones are in each pile?

_____

8. Tanner has 30 stickers. He puts 6 stickers on each page. On how many pages does he put stickers?

_____

© Houghton Mifflin Harcourt Publishing Company

Lesson Check

1. Jack and his little sister are stacking 24 blocks. They put the blocks in 3 equal stacks. How many blocks are in each stack?

   (A) 4

   (B) 6

   (C) 7

   (D) 8

2. Melissa made 45 greeting cards. She put them in 5 equal piles. How many cards did she put in each pile?

   (A) 9

   (B) 8

   (C) 7

   (D) 6

Spiral Review

3. Angie puts 1 stamp on each envelope. She puts stamps on 7 envelopes. How many stamps does Angie use? **(Lesson 3.7)**

   (A) 0

   (B) 1

   (C) 7

   (D) 8

4. A carnival ride has 8 cars. Each car holds 4 people. How many people are on the ride if all the cars are full? **(Lesson 4.8)**

   (A) 34

   (B) 32

   (C) 28

   (D) 24

Use the line plot for 5–6.

5. How many families have 1 computer at home? **(Lesson 2.7)**

   (A) 4   (C) 6

   (B) 5   (D) 7

6. How many families have more than 1 computer at home? **(Lesson 2.7)**

   (A) 4   (C) 7

   (B) 5   (D) 8

```
              X
        X     X
        X     X           X
        X     X     X     X
        X     X     X     X
        X     X     X     X     X
        +-----+-----+-----+-----+
        0     1     2     3     4
```

**Number of
Computers at Home**

© Houghton Mifflin Harcourt Publishing Company

Name _____

Relate Subtraction and Division

Write a division equation.

1.
$$\begin{array}{ccccccc} 16 & & 12 & & 8 & & 4 \\ -\ 4 & & -\ 4 & & -\ 4 & & -\ 4 \\ \hline 12 & & 8 & & 4 & & 0 \end{array}$$

$$16 \div 4 = 4$$

2.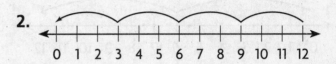

_____

3.

_____

4.
$$\begin{array}{ccccccc} 20 & & 15 & & 10 & & 5 \\ -\ 5 & & -\ 5 & & -\ 5 & & -\ 5 \\ \hline 15 & & 10 & & 5 & & 0 \end{array}$$

_____

Use repeated subtraction or a number line to solve.

5. $28 \div 7 =$ _____

6. $18 \div 6 =$ _____

7. $8\overline{)40}$

8. $9\overline{)36}$

Problem Solving REAL WORLD

9. Mrs. Costa has 18 pencils. She gives 9 pencils to each of her children for school. How many children does Mrs. Costa have?

10. Boël decides to plant rose bushes in her garden. She has 24 bushes. She places 6 bushes in each row. How many rows of rose bushes does she plant in her garden?

_____

_____

© Houghton Mifflin Harcourt Publishing Company

Lesson Check

1. Which division equation is shown?

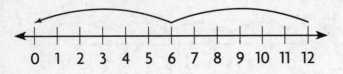

   Ⓐ $3 \times 4 = 12$ Ⓒ $12 \div 3 = 4$

   Ⓑ $12 \div 6 = 2$ Ⓓ $12 \div 4 = 3$

2. Isabella has 35 cups of dog food. She feeds her dogs 5 cups of food each day. For how many days will the dog food last?

   Ⓐ 6 days   Ⓒ 8 days

   Ⓑ 7 days   Ⓓ 9 days

Spiral Review

3. Ellen buys 4 bags of oranges. There are 6 oranges in each bag. How many oranges does Ellen buy?

(Lesson 4.3)

   Ⓐ 10   Ⓒ 24

   Ⓑ 12   Ⓓ 30

4. Each month for 7 months, Samuel mows 3 lawns. How many more lawns does he need to mow before he has mowed 29 lawns? (Lesson 4.10)

   Ⓐ 1   Ⓒ 7

   Ⓑ 3   Ⓓ 8

Use the graph for 5–6.

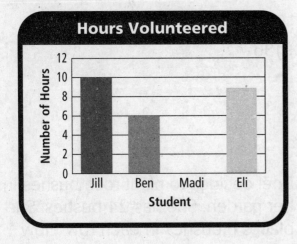

5. How many hours did Eli volunteer?

(Lesson 2.4)

   Ⓐ 4 hours   Ⓒ 9 hours

   Ⓑ 8 hours   Ⓓ 10 hours

6. Madi volunteered 2 hours less than Jill. At what number should the bar for Madi end? (Lesson 2.5)

   Ⓐ 3   Ⓒ 8

   Ⓑ 6   Ⓓ 12

© Houghton Mifflin Harcourt Publishing Company

Model with Arrays

Use square tiles to make an array. Solve.

1. How many rows of 4 are in 12?

_____3 rows_____

2. How many rows of 3 are in 21?

_____

3. How many rows of 6 are in 30?

_____

4. How many rows of 9 are in 18?

_____

Make an array. Then write a division equation.

5. 20 tiles in 5 rows

_____

6. 28 tiles in 7 rows

_____

7. 18 tiles in 9 rows

_____

8. 36 tiles in 6 rows

_____

Problem Solving REAL WORLD

9. A dressmaker has 24 buttons. He needs 3 buttons to make one dress. How many dresses can he make with 24 buttons?

_____

10. Liana buys 36 party favors for her 9 guests. She gives an equal number of favors to each guest. How many party favors does each guest get?

_____

© Houghton Mifflin Harcourt Publishing Company

Lesson Check

1. Mr. Canton places 24 desks in 6 equal rows. How many desks are in each row?

(A) 2

(B) 3

(C) 4

(D) 5

2. Which division equation is shown by the array?

(A) $12 \div 6 = 2$ (C) $12 \div 2 = 6$

(B) $12 \div 3 = 4$ (D) $12 \div 1 = 12$

Spiral Review

3. Amy has 2 rows of 4 sports trophies on each of her 3 shelves. How many sports trophies does Amy have in all? **(Lesson 4.6)**

(A) 8

(B) 9

(C) 12

(D) 24

4. What is the unknown factor?

(Lesson 5.2)

$9 \times p = 45$

(A) 4

(B) 5

(C) 6

(D) 7

5. Sam has 7 stacks with 4 quarters each. How many quarters does Sam have? **(Lesson 4.5)**

(A) 11

(B) 12

(C) 24

(D) 28

6. How can you skip count to find how many counters in all? **(Lesson 3.1)**

(A) 3 groups of 2

(B) 3 groups of 3

(C) 9 groups of 2

(D) 18 groups of 2

© Houghton Mifflin Harcourt Publishing Company

Relate Multiplication and Division

Complete the equations.

1.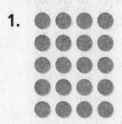

   5 rows of __4__ = 20

   5 × __4__ = 20

   20 ÷ 5 = __4__

2. (dots array)

   4 rows of _____ = 24

   4 × _____ = 24

   24 ÷ 4 = _____

3. (dots array)

   3 rows of _____ = 24

   3 × _____ = 24

   24 ÷ 3 = _____

Complete the equations.

4. 4 × _____ = 28   28 ÷ 4 = _____

5. 6 × _____ = 36   36 ÷ 6 = _____

6. 7 × _____ = 35   35 ÷ 7 = _____

7. 7 × _____ = 21   21 ÷ 7 = _____

8. 9 × _____ = 27   27 ÷ 9 = _____

9. 2 × _____ = 16   16 ÷ 2 = _____

10. 4 × _____ = 36   36 ÷ 4 = _____

11. 8 × _____ = 40   40 ÷ 8 = _____

Problem Solving

12. Mr. Martin buys 36 muffins for a class breakfast. He places them on plates for his students. If he places 9 muffins on each plate, how many plates does Mr. Martin use?

13. Ralph read 18 books during his summer vacation. He read the same number of books each month for 3 months. How many books did he read each month?

_____   _____

© Houghton Mifflin Harcourt Publishing Company

Lesson Check

1. Which number will complete the equations?

    $6 \times \blacksquare = 24$

    $24 \div 6 = \blacksquare$

    (A) 3   (C) 5

    (B) 4   (D) 6

2. Alice has 14 seashells. She divides them equally between her 2 sisters. How many seashells does each sister get?

    (A) 7   (C) 12

    (B) 8   (D) 16

Spiral Review

3. Sam and Jesse can each wash 5 cars in an hour. They both work for 7 hours over 2 days. How many cars did Sam and Jesse wash?

    (Lesson 4.6)

    (A) 70

    (B) 35

    (C) 24

    (D) 14

4. Keisha skip counted to find how many counters in all. How many equal groups are there? (Lesson 3.1)

    _____ groups of 5

    (A) 3   (C) 5

    (B) 4   (D) 20

5. The key for a picture graph showing the number of books students read is: Each  = 2 books. How many books did Nancy read if she has ▯▯▮ by her name? (Lesson 2.2)

    (A) 2

    (B) 4

    (C) 5

    (D) 6

6. Jan surveyed her friends to find their favorite season. She recorded ЖⅢ Ⅲ for summer. How many people chose summer as their favorite season? (Lesson 2.1)

    (A) 5

    (B) 8

    (C) 9

    (D) 13

© Houghton Mifflin Harcourt Publishing Company

Name _____

Write Related Facts

Write the related facts for the array.

1.
$$2 \times 6 = 12$$
$$6 \times 2 = 12$$
$$12 \div 2 = 6$$
$$12 \div 6 = 2$$

2. _____
_____
_____
_____

3. _____
_____
_____
_____

Write the related facts for the set of numbers.

4. 3, 7, 21

_____
_____
_____
_____

5. 2, 9, 18

_____
_____
_____
_____

6. 4, 8, 32

_____
_____
_____
_____

Complete the related facts.

7. $4 \times 9 =$ _____

$9 \times$ _____ $= 36$

$36 \div$ _____ $= 4$

_____ $\div 4 = 9$

8. _____ $\times 7 = 35$

_____ $\times 5 = 35$

_____ $\div 7 = 5$

$35 \div 5 =$ _____

9. $6 \times$ _____ $= 18$

$3 \times 6 =$ _____

$18 \div$ _____ $= 3$

_____ $\div 3 = 6$

Problem Solving REAL WORLD

10. CDs are on sale for $5 each. Jennifer has $45 and wants to buy as many as she can. How many CDs can Jennifer buy?

_____

11. Mr. Moore has 21 feet of wallpaper. He cuts it into sections that are each 3 feet long. How many sections does Mr. Moore have?

_____

© Houghton Mifflin Harcourt Publishing Company

Lesson Check

1. Which number completes the set of related facts?

   $5 \times \blacksquare = 40$ $40 \div \blacksquare = 5$

   $\blacksquare \times 5 = 40$ $40 \div 5 = \blacksquare$

   (A) 6

   (B) 7

   (C) 8

   (D) 9

2. Which equation is not in the same set of related facts as $4 \times 7 = 28$?

   (A) $7 \times 4 = 28$

   (B) $4 + 7 = 11$

   (C) $28 \div 4 = 7$

   (D) $28 \div 7 = 4$

Spiral Review

3. Beth runs 20 miles each week for 8 weeks. How many miles does Beth run in 8 weeks? (Lesson 5.5)

   (A) 16 miles

   (B) 28 miles

   (C) 100 miles

   (D) 160 miles

4. Find the product. (Lesson 3.7)

   $$5 \times 0$$

   (A) 0

   (B) 1

   (C) 5

   (D) 10

5. Uri's bookcase has 5 shelves. There are 9 books on each shelf. How many books in all are in Uri's bookcase? (Lesson 4.9)

   (A) 14

   (B) 36

   (C) 45

   (D) 54

6. There are 6 batteries in one package. How many batteries will 6 packages have? (Lesson 3.1)

   (A) 12

   (B) 18

   (C) 24

   (D) 36

© Houghton Mifflin Harcourt Publishing Company

Name _____

Division Rules for 1 and 0

Find the quotient.

1. $3 \div 1 =$ __3__   **2.** $8 \div 8 =$ _____   **3.** _____ $= 0 \div 6$ **4.** $2 \div 2 =$ _____

5. _____ $= 9 \div 1$ **6.** $0 \div 2 =$ _____   **7.** $0 \div 3 =$ _____   **8.** _____ $= 0 \div 4$

9. $7\overline{)7}$ **10.** $1\overline{)6}$ **11.** $9\overline{)0}$ **12.** $1\overline{)5}$

13. $1\overline{)0}$ **14.** $4\overline{)4}$ **15.** $1\overline{)10}$ **16.** $2\overline{)2}$

Problem Solving  REAL WORLD

17. There are no horses in the stables. There are 3 stables in all. How many horses are in each stable?

_____

18. Jon has 6 kites. He and his friends will each fly 1 kite. How many people in all will fly a kite?

_____

© Houghton Mifflin Harcourt Publishing Company

Lesson Check

1. Candace has 6 pairs of jeans. She places each pair on its own hanger. How many hangers does Candace use?

- Ⓐ 0
- Ⓒ 6
- Ⓑ 1
- Ⓓ 12

2. There are 0 birds and 4 bird cages. Which division equation describes how many birds are in each cage?

- Ⓐ $0 \div 4 = 0$
- Ⓒ $4 \div 1 = 4$
- Ⓑ $4 \div 4 = 1$
- Ⓓ $0 \times 4 = 0$

Spiral Review

3. There are 7 plates on the table. There are 0 sandwiches on each plate. How many sandwiches are on the plates in all? **(Lesson 3.7)**

7×0

- Ⓐ 0
- Ⓑ 1
- Ⓒ 7
- Ⓓ 70

4. Which shows a way to break apart the array to find the product? **(Lesson 4.4)**

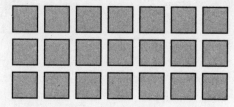

- Ⓐ $(3 \times 5) + (3 \times 2)$
- Ⓑ $(2 \times 8) + (1 \times 8)$
- Ⓒ $(4 \times 7) + (1 \times 7)$
- Ⓓ $(3 \times 6) + (3 \times 3)$

5. Which of the following describes a pattern in the table? **(Lesson 5.1)**

Vans	1	2	3	4	5
Students	6	12	18	24	30

- Ⓐ Add 5.
- Ⓑ Multiply by 2.
- Ⓒ Subtract 1.
- Ⓓ Multiply by 6.

6. Use the graph.

How many more cans did Sam bring in than Lee? **(Lesson 2.5)**

- Ⓐ 4
- Ⓒ 7
- Ⓑ 5
- Ⓓ 9

© Houghton Mifflin Harcourt Publishing Company

Chapter 6 Extra Practice

Lessons 6.1 - 6.3

Make equal groups. Complete the table.

	Counters	Number of Equal Groups	Number in Each Group
1.	18	9	
2.	24		8
3.	12	6	
4.	35	7	
5.	32		4
6.	25		5

Lesson 6.4

Write a division equation for the picture.

1.

2.

_____

_____

_____

_____

Lesson 6.5

Write a division equation.

1.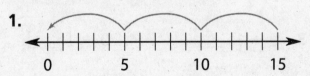

$$\begin{array}{c} 0 \quad\quad 5 \quad\quad 10 \quad\quad 15 \end{array}$$

2.
$$\begin{array}{cccc} 24 & 18 & 12 & 6 \\ -\ 6 & -\ 6 & -\ 6 & -\ 6 \\ \hline 18 & 12 & 6 & 0 \end{array}$$

_____

_____

© Houghton Mifflin Harcourt Publishing Company

Lesson 6.6

Make an array. Then write a division equation.

1. 12 tiles in 4 rows

_____

2. 18 tiles in 3 rows

_____

3. 35 tiles in 5 rows

_____

4. 28 tiles in 7 rows

_____

Lesson 6.7

Complete the equations.

1. $8 \times$ ___ $= 40$ $40 \div 8 =$ ___

2. $6 \times$ ___ $= 36$ $36 \div 6 =$ ___

3. $3 \times$ ___ $= 21$ $21 \div 3 =$ ___

4. $2 \times$ ___ $= 18$ $18 \div 2 =$ ___

Lesson 6.8 (pp. 239–243)

Write the related facts for the array.

1.

2.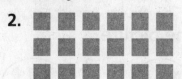

3.

_____   _____   _____

_____   _____   _____

_____   _____   _____

Lesson 6.9

Find the quotient.

1. $7 \div 1$ ___

2. $4 \div 4$ ___

3. $9 \div 1$ ___

4. $0 \div 1$ ___

5. Anton has 8 flower pots. He plants 1 seed in each pot. How many seeds does Anton use?

_____

© Houghton Mifflin Harcourt Publishing Company

School-Home Letter

© Houghton Mifflin Harcourt Publishing Company

Vocabulary

array An arrangement of objects in rows and columns

equation A number sentence that uses the equal sign to show that two amounts are equal

order of operations A special set of rules that gives the order in which calculations are done to solve a problem

related facts A set of related multiplication and division equations

Dear Family,

During the next few weeks, our math class will be learning about division facts and strategies. We will learn strategies to use to divide by 2, 3, 4, 5, 6, 7, 8, 9, and 10. We will also learn the order of operations rules to solve problems involving more than one operation.

You can expect to see homework that provides practice with dividing by these divisors.

Here is a sample of how your child will be taught to divide.

🔒 MODEL Use an Array

This is how we can use arrays to divide.

STEP 1

$20 \div 4 = $ ▩

Draw rows of 4 tiles until you have drawn all 20 tiles.

□ □ □ □
□ □ □ □
□ □ □ □
□ □ □ □
□ □ □ □

STEP 2

Count the number of rows to find the quotient.

There are 5 rows of 4 tiles.

So, $20 \div 4 = 5$.

Tips

Use a Related Multiplication Fact

Since division is the opposite of multiplication, using a multiplication fact is another way to find a quotient. To divide 20 by 4, for example, think of a related multiplication fact: $4 \times $ ▩ $ = 20$.
$4 \times 5 = 20$.
So, $20 \div 4 = 5$.

Activity

Provide 12 pennies. Have your child make as many arrays as possible using all 12 pennies. Have your child write a division equation for each array.

Carta para la casa

Vocabulario

ecuación Una oración numérica que usa el signo de igual para mostrar que dos cantidades son iguales

matriz Una forma de ordenar objetos en filas y columnas

orden de las operaciones Un conjunto especial de reglas que expresa el orden en el que se realizan las operaciones para resolver un problema

Querida familia,

Durante las próximas semanas, en la clase de matemáticas aprenderemos sobre las operaciones de división y sus estrategias. Aprenderemos estrategias para dividir entre 2, 3, 4, 5, 6, 7, 8, 9 y 10. También aprenderemos las reglas del orden de las operaciones para resolver problemas en los que hay más de una operación.

Llevaré a la casa tareas que sirven para practicar la división entre estos divisores.

Este es un ejemplo de la manera como aprenderemos a dividir.

🔑 MODELO Usar una matriz

Esta es la manera como podemos usar matrices para dividir.

Pistas

Usar una operación de multiplicación relacionada

Dado que la división es opuesta a la multiplicación, usar una operación de multiplicación es otra manera de hallar un cociente. Para dividir 20 entre 4, por ejemplo, piensa en una operación de multiplicación relacionada: $4 \times$ $= 20$. $4 \times 5 = 20$. Por lo tanto, $20 \div 4 = 5$.

PASO 1

$20 \div 4 =$ ▪

Traza filas de 4 fichas cuadradas hasta tener las 20 fichas.

☐ ☐ ☐ ☐
☐ ☐ ☐ ☐
☐ ☐ ☐ ☐
☐ ☐ ☐ ☐
☐ ☐ ☐ ☐

PASO 2

Cuenta la cantidad de filas para encontrar el cociente.

Hay 5 filas de 4 fichas.

Por lo tanto, $20 \div 4 = 5$.

Actividad

Dé a su hijo 12 monedas de 1¢. Pídale que haga la mayor cantidad posible de matrices usando las 12 monedas de 1¢. Luego, pídale que escriba un enunciado de división para cada matriz.

© Houghton Mifflin Harcourt Publishing Company

Divide by 2

Write a division equation for the picture.

1.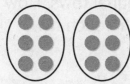

2.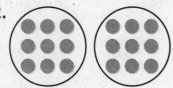

3.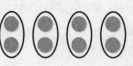

$$12 \div 2 = 6 \text{ or}$$
$$\underline{12 \div 6 = 2}$$

_____

_____

_____

_____

Find the quotient. You may want to draw a quick picture to help.

4. ____ $= 14 \div 2$

5. ____ $= 4 \div 2$

6. $16 \div 2 =$ ____

7. $2\overline{)18}$

8. $2\overline{)12}$

9. $2\overline{)14}$

Problem Solving

10. Mr. Reynolds, the gym teacher, divided a class of 16 students into 2 equal teams. How many students were on each team?

_____

11. Sandra has 10 books. She divides them into groups of 2 each. How many groups can she make?

_____

© Houghton Mifflin Harcourt Publishing Company

Lesson Check

1. Ava has 12 apples and 2 baskets. She puts an equal number of apples in each basket. How many apples are in a basket?

   (A) 2

   (B) 4

   (C) 6

   (D) 8

2. There are 8 students singing a song in the school musical. Ms. Lang put the students in 2 equal rows. How many students are in each row?

   (A) 2

   (B) 4

   (C) 6

   (D) 10

Spiral Review

3. Find the product. (Lesson 4.1)

   2×6

   (A) 4

   (B) 8

   (C) 12

   (D) 18

4. Jayden plants 24 trees. He plants the trees equally in 3 rows. How many trees are in each row?

   (Lesson 6.2)

   (A) 6

   (B) 8

   (C) 9

   (D) 27

5. Which of the following describes this pattern? (Lesson 4.7)

   9, 12, 15, 18, 21, 24

   (A) Multiply by 3.

   (B) Multiply by 5.

   (C) Add 3.

   (D) Subtract 3.

6. A tricycle has 3 wheels. How many wheels are there on 4 tricycles?

   (Lesson 4.3)

   (A) 7

   (B) 9

   (C) 12

   (D) 15

© Houghton Mifflin Harcourt Publishing Company

Name _____

Divide by 10

Find the unknown factor and quotient.

1. $10 \times \underline{\mathbf{2}} = 20$ $20 \div 10 = \underline{\mathbf{2}}$

2. $10 \times \underline{\hspace{1cm}} = 70$ $70 \div 10 = \underline{\hspace{1cm}}$

3. $10 \times \underline{\hspace{1cm}} = 80$ $80 \div 10 = \underline{\hspace{1cm}}$

4. $10 \times \underline{\hspace{1cm}} = 30$ $30 \div 10 = \underline{\hspace{1cm}}$

Find the quotient.

5. $60 \div 10 = \underline{\hspace{1cm}}$

6. $\underline{\hspace{1cm}} = 40 \div 4$

7. $20 \div 2 = \underline{\hspace{1cm}}$

8. $50 \div 10 = \underline{\hspace{1cm}}$

9. $90 \div 10 = \underline{\hspace{1cm}}$

10. $10 \div 10 = \underline{\hspace{1cm}}$

11. $\underline{\hspace{1cm}} = 30 \div 10$

12. $40 \div 10 = \underline{\hspace{1cm}}$

13. $10\overline{)40}$

14. $10\overline{)70}$

15. $10\overline{)100}$

16. $10\overline{)20}$

Problem Solving 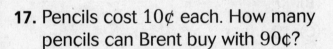 REAL WORLD

17. Pencils cost 10¢ each. How many pencils can Brent buy with 90¢?

_____

18. Mrs. Marks wants to buy 80 pens. If the pens come in packs of 10, how many packs does she need to buy?

_____

© Houghton Mifflin Harcourt Publishing Company

Lesson Check

1. Gracie uses 10 beads on each necklace she makes. She has 60 beads to use. How many necklaces can Gracie make?

Ⓐ 6   Ⓒ 50

Ⓑ 10   Ⓓ 70

2. A florist arranges 10 flowers in each vase. How many vases does the florist need to arrange 40 flowers?

Ⓐ 3   Ⓒ 30

Ⓑ 4   Ⓓ 50

Spiral Review

3. What is the unknown factor?

(Lesson 5.2)

$7 \times p = 14$

Ⓐ 21

Ⓑ 7

Ⓒ 3

Ⓓ 2

4. Aspen Bakery sold 40 boxes of rolls in one day. Each box holds 6 rolls. How many rolls in all did the bakery sell? (Lesson 5.4)

Ⓐ 24

Ⓑ 46

Ⓒ 240

Ⓓ 320

5. Mr. Samuels buys a sheet of stamps. There are 4 rows with 7 stamps in each row. How many stamps does Mr. Samuels buy?

(Lesson 4.1)

Ⓐ 11

Ⓑ 14

Ⓒ 21

Ⓓ 28

6. There are 56 students going on a field trip to the science center. The students tour the center in groups of 8. How many groups of students are there? (Lesson 6.2)

Ⓐ 6

Ⓑ 7

Ⓒ 9

Ⓓ 64

© Houghton Mifflin Harcourt Publishing Company

Name _____

Divide by 5

Use count up or count back on a number line to solve.

1. $40 \div 5 =$ __**8**__

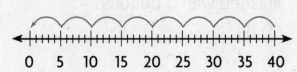

0 5 10 15 20 25 30 35 40

2. $25 \div 5 =$ ____

0 5 10 15 20 25

Find the quotient.

3. ____ $= 10 \div 5$ **4.** ____ $= 30 \div 5$ **5.** $14 \div 2 =$ ____   **6.** $5 \div 5 =$ ____

7. $45 \div 5 =$ ____   **8.** ____ $= 60 \div 10$ **9.** ____ $= 15 \div 5$ **10.** $18 \div 2 =$ ____

11. ____ $= 0 \div 5$ **12.** $20 \div 5 =$ ____   **13.** $25 \div 5 =$ ____   **14.** ____ $= 35 \div 5$

15. $5\overline{)20}$ **16.** $10\overline{)70}$ **17.** $5\overline{)15}$ **18.** $5\overline{)40}$

Problem Solving

19. A model car maker puts 5 wheels in each kit. A machine makes 30 wheels at a time. How many packages of 5 wheels can be made from the 30 wheels?

20. A doll maker puts a small bag with 5 hair ribbons inside each box with a doll. How many bags of 5 hair ribbons can be made from 45 hair ribbons?

_____   _____

© Houghton Mifflin Harcourt Publishing Company

Lesson Check

1. A model train company puts 5 boxcars with each train set. How many sets can be completed using 35 boxcars?

   (A) 5

   (B) 6

   (C) 7

   (D) 8

2. A machine makes 5 buttons at a time. Each doll shirt gets 5 buttons. How many doll shirts can be finished with 5 buttons?

   (A) 0

   (B) 1

   (C) 2

   (D) 5

Spiral Review

3. Julia earns $5 each day running errands for a neighbor. How much will Julia earn if she runs errands for 6 days in one month? (Lesson 4.3)

   (A) $40

   (B) $35

   (C) $30

   (D) $25

4. Marcus has 12 slices of bread. He uses 2 slices of bread for each sandwich. How many sandwiches can Marcus make? (Lesson 7.1)

   (A) 6

   (B) 7

   (C) 8

   (D) 9

Use the line plot for 5–6.

5. How many students have no pets? (Lesson 2.7)

   (A) 0   (C) 4

   (B) 3   (D) 5

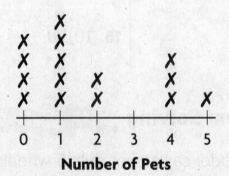

Number of Pets

6. How many students answered the question "How many pets do you have?" (Lesson 2.7)

   (A) 10   (C) 14

   (B) 12   (D) 15

© Houghton Mifflin Harcourt Publishing Company

Divide by 3

Find the quotient. Draw a quick picture to help.

1. $12 ÷ 3 = \underline{4}$

2. $24 ÷ 3 = \underline{}$

3. $\underline{} = 6 ÷ 3$

4. $40 ÷ 5 = \underline{}$

Find the quotient.

5. $\underline{} = 15 ÷ 3$

6. $\underline{} = 21 ÷ 3$

7. $16 ÷ 2 = \underline{}$

8. $27 ÷ 3 = \underline{}$

9. $0 ÷ 3 = \underline{}$

10. $9 ÷ 3 = \underline{}$

11. $\underline{} = 30 ÷ 3$

12. $\underline{} = 12 ÷ 4$

13. $3\overline{)12}$

14. $3\overline{)15}$

15. $3\overline{)24}$

16. $3\overline{)9}$

Problem Solving REAL WORLD

17. The principal at Miller Street School has 12 packs of new pencils. She will give 3 packs to each third-grade class. How many third-grade classes are there?

_____

18. Mike has $21 to spend at the mall. He spends all of his money on bracelets for his sisters. Bracelets cost $3 each. How many bracelets does he buy?

_____

© Houghton Mifflin Harcourt Publishing Company

Lesson Check

1. There are 18 counters divided equally among 3 groups. How many counters are in each group?

Ⓐ 5

Ⓑ 6

Ⓒ 7

Ⓓ 8

2. Josh has 27 signed baseballs. He places the baseballs equally on 3 shelves. How many baseballs are on each shelf?

Ⓐ 6

Ⓑ 7

Ⓒ 8

Ⓓ 9

Spiral Review

3. Each bicycle has 2 wheels. How many wheels do 8 bicycles have?

(Lesson 3.1)

Ⓐ 10

Ⓑ 16

Ⓒ 24

Ⓓ 32

4. How many students watch less than 3 hours of TV a day? (Lesson 2.7)

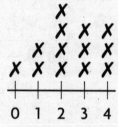

Hours Watching TV

Ⓐ 3   Ⓒ 8

Ⓑ 7   Ⓓ 13

5. Which of the following is an example of the Distributive Property? (Lesson 4.4)

Ⓐ $3 \times 6 = 18$

Ⓑ $6 \times 3 = 15 + 3$

Ⓒ $3 \times 6 = 6 \times 3$

Ⓓ $3 \times 6 = (3 \times 2) + (3 \times 4)$

6. Which unknown number completes the equations? (Lesson 6.7)

$$3 \times \blacksquare = 21 \qquad 21 \div 3 = \blacksquare$$

Ⓐ 3

Ⓑ 6

Ⓒ 7

Ⓓ 18

© Houghton Mifflin Harcourt Publishing Company

Name _____

Divide by 4

Draw tiles to make an array. Find the quotient.

1. ___4___ = 16 ÷ 4

[array of tiles, 4 rows of 4]

2. 20 ÷ 4 = _____

3. 12 ÷ 4 = _____

4. 10 ÷ 2 = _____

Find the quotient.

5. 24 ÷ 3 = _____

6. _____ = 8 ÷ 2

7. 32 ÷ 4 = _____

8. _____ = 28 ÷ 4

9. $4\overline{)36}$

10. $4\overline{)8}$

11. $4\overline{)24}$

12. $3\overline{)30}$

Find the unknown number.

13. 20 ÷ 5 = a

14. 32 ÷ 4 = p

15. 40 ÷ 10 = ■

16. 18 ÷ 3 = x

a = _____

p = _____

■ = _____

x = _____

Problem Solving  REAL WORLD

17. Ms. Higgins has 28 students in her gym class. She puts them in 4 equal groups. How many students are in each group?

_____

18. Andy has 36 CDs. He buys a case that holds 4 CDs in each section. How many sections can he fill?

_____

© Houghton Mifflin Harcourt Publishing Company

Lesson Check

1. Darion picks 16 grapefruits off a tree in his backyard. He puts 4 grapefruits in each bag. How many bags does he need?

Ⓐ 3

Ⓑ 4

Ⓒ 5

Ⓓ 6

2. Tori has a bag of 32 markers to share equally among 3 friends and herself. How many markers will Tori and each of her friends get?

Ⓐ 6

Ⓑ 7

Ⓒ 8

Ⓓ 9

Spiral Review

3. Find the product. (Lesson 4.5)

3×7

Ⓐ 18

Ⓑ 21

Ⓒ 24

Ⓓ 28

4. Which of the following describes this pattern? (Lesson 4.7)

8, 12, 16, 20, 24, 28

Ⓐ Multiply by 4.

Ⓑ Add 4.

Ⓒ Multiply by 2.

Ⓓ Subtract 4.

5. Which is an example of the Commutative Property of Multiplication? (Lesson 3.6)

Ⓐ $3 \times 6 = 2 \times 9$

Ⓑ $2 \times 4 = 5 + 3$

Ⓒ $4 \times 5 = 5 \times 4$

Ⓓ $2 \times 5 = 5 + 5$

6. Jasmine has 18 model horses. She places the model horses equally on 3 shelves. How many model horses are on each shelf? (Lesson 6.2)

Ⓐ 6

Ⓑ 7

Ⓒ 15

Ⓓ 21

© Houghton Mifflin Harcourt Publishing Company

Name _____

Divide by 6

Find the unknown factor and quotient.

1. $6 \times \underline{7} = 42$ $42 \div 6 = \underline{7}$

2. $6 \times \underline{\quad} = 18$ $18 \div 6 = \underline{\quad}$

3. $4 \times \underline{\quad} = 24$ $24 \div 4 = \underline{\quad}$

4. $6 \times \underline{\quad} = 54$ $54 \div 6 = \underline{\quad}$

Find the quotient.

5. $\underline{\quad} = 24 \div 6$

6. $48 \div 6 = \underline{\quad}$

7. $\underline{\quad} = 6 \div 6$

8. $12 \div 6 = \underline{\quad}$

9. $6\overline{)36}$

10. $6\overline{)54}$

11. $6\overline{)30}$

12. $1\overline{)6}$

Find the unknown number.

13. $p = 42 \div 6$

14. $18 \div 3 = q$

15. $r = 30 \div 6$

16. $60 \div 6 = s$

$p = \underline{\quad}$

$q = \underline{\quad}$

$r = \underline{\quad}$

$s = \underline{\quad}$

Problem Solving REAL WORLD

17. Lucas has 36 pages of a book left to read. If he reads 6 pages a day, how many days will it take Lucas to finish the book?

_____

18. Juan has $24 to spend at the bookstore. If books cost $6 each, how many books can he buy?

_____

© Houghton Mifflin Harcourt Publishing Company

Lesson Check

1. Ella earned $54 last week babysitting. She earns $6 an hour. How many hours did Ella babysit last week?

   (A) 6 hours

   (B) 7 hours

   (C) 8 hours

   (D) 9 hours

2. What is the unknown factor and quotient?

   $6 \times \blacksquare = 42$ $42 \div 6 = \blacksquare$

   (A) 6

   (B) 7

   (C) 8

   (D) 9

Spiral Review

3. Coach Clarke has 48 students in his P.E. class. He places the students in teams of 6 for an activity. How many teams can Coach Clarke make? (Lesson 6.3)

   (A) 7

   (B) 8

   (C) 9

   (D) 54

4. Each month for 7 months, Eva reads 3 books. How many more books does she need to read before she has read 30 books?

   (Lesson 4.10)

   (A) 7

   (B) 9

   (C) 27

   (D) 33

5. Each cow has 4 legs. How many legs will 5 cows have? (Lesson 3.1)

   (A) 9

   (B) 16

   (C) 20

   (D) 24

6. Find the product. (Lesson 4.9)

   3×9

   (A) 36

   (B) 27

   (C) 18

   (D) 12

© Houghton Mifflin Harcourt Publishing Company

Name _____

Divide by 7

Find the unknown factor and quotient.

1. $7 \times \underline{6} = 42 \quad 42 \div 7 = \underline{6}$ | 2. $7 \times \underline{} = 35 \quad 35 \div 7 = \underline{}$

3. $7 \times \underline{} = 7 \quad 7 \div 7 = \underline{}$ | 4. $5 \times \underline{} = 20 \quad 20 \div 5 = \underline{}$

Find the quotient.

5. $7\overline{)21}$ 6. $7\overline{)14}$ 7. $6\overline{)48}$ 8. $7\overline{)63}$

9. $\underline{} = 35 \div 7$ 10. $0 \div 7 = \underline{}$ 11. $\underline{} = 56 \div 7$ 12. $32 \div 8 = \underline{}$

Find the unknown number.

13. $56 \div 7 = e$ 14. $k = 32 \div 4$ 15. $g = 49 \div 7$ 16. $28 \div 7 = s$

    $e = \underline{}$ $k = \underline{}$ $g = \underline{}$ $s = \underline{}$

Problem Solving  REAL WORLD

17. Twenty-eight players sign up for basketball. The coach puts 7 players on each team. How many teams are there?

18. Roberto read 42 books over 7 months. He read the same number of books each month. How many books did Roberto read each month?

_____

_____

© Houghton Mifflin Harcourt Publishing Company

Lesson Check

1. Elliot earned $49 last month walking his neighbor's dog. He earns $7 each time he walks the dog. How many times did Elliot walk his neighbor's dog last month?

Ⓐ 6   Ⓒ 8

Ⓑ 7   Ⓓ 9

2. Which is the unknown factor and quotient?

$7 \times \blacksquare = 63$

$63 \div 7 = \blacksquare$

Ⓐ 6   Ⓒ 8

Ⓑ 7   Ⓓ 9

Spiral Review

3. Maria puts 6 strawberries in each smoothie she makes. She makes 3 smoothies. Altogether, how many strawberries does Maria use in the smoothies? **(Lesson 4.3)**

Ⓐ 9

Ⓑ 12

Ⓒ 18

Ⓓ 24

4. Kaitlyn makes 4 bracelets. She uses 8 beads for each bracelet. How many beads does she use in all?

(Lesson 4.8)

Ⓐ 12

Ⓑ 16

Ⓒ 32

Ⓓ 40

5. What is the unknown factor?

(Lesson 3.6)

$2 \times 5 = 5 \times \blacksquare$

Ⓐ 10

Ⓑ 5

Ⓒ 2

Ⓓ 1

6. Which division equation is related to the following multiplication equation? **(Lesson 6.7)**

$3 \times 4 = 12$

Ⓐ $12 \div 4 = 3$

Ⓑ $8 \div 2 = 4$

Ⓒ $12 \div 2 = 6$

Ⓓ $10 \div 5 = 2$

© Houghton Mifflin Harcourt Publishing Company

Name _____

Divide by 8

Find the unknown factor and quotient.

1. $8 \times$ __4__ $= 32$ $32 \div 8 =$ ____

2. $3 \times$ ____ $= 27$ $27 \div 3 =$ ____

3. $8 \times$ ____ $= 8$ $8 \div 8 =$ ____

4. $8 \times$ ____ $= 72$ $72 \div 8 =$ ____

Find the quotient.

5. ____ $= 24 \div 8$ **6.** $40 \div 8 =$ ____   **7.** ____ $= 56 \div 8$ **8.** $14 \div 2 =$ ____

9. $8\overline{)64}$ **10.** $7\overline{)28}$ **11.** $8\overline{)16}$ **12.** $8\overline{)48}$

Find the unknown number.

13. $16 \div p = 8$ **14.** $25 \div \blacksquare = 5$ **15.** $24 \div a = 3$ **16.** $k \div 10 = 8$

$p =$ ____ $\blacksquare =$ ____ $a =$ ____ $k =$ ____

Problem Solving  REAL WORLD

17. Sixty-four students are going on a field trip. There is 1 adult for every 8 students. How many adults are there?

_____

18. Mr. Chen spends $32 for tickets to a play. If the tickets cost $8 each, how many tickets does Mr. Chen buy?

_____

© Houghton Mifflin Harcourt Publishing Company

Lesson Check

1. Mrs. Wilke spends $72 on pies for the school fair. Each pie costs $8. How many pies does Mrs. Wilke buy for the school fair?

    (A) 6
    (B) 7
    (C) 8
    (D) 9

2. Find the unknown factor and quotient.

    $8 \times \blacksquare = 40$

    $40 \div 8 = \blacksquare$

    (A) 4
    (B) 5
    (C) 6
    (D) 7

Spiral Review

3. Find the product. (Lesson 4.6)

    $(3 \times 2) \times 5$

    (A) 6
    (B) 10
    (C) 20
    (D) 30

4. Which of the following has the same product as 4×9? (Lesson 3.6)

    (A) 3×8
    (B) 9×4
    (C) 5×6
    (D) 7×2

5. Find the unknown factor. (Lesson 5.2)

    $8 \times \blacksquare = 32$

    (A) 4
    (B) 5
    (C) 6
    (D) 24

6. Which multiplication sentence represents the array? (Lesson 3.5)

    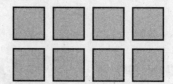

    (A) $1 \times 8 = 8$
    (B) $4 + 4 = 8$
    (C) $2 \times 4 = 8$
    (D) $4 \times 3 = 12$

© Houghton Mifflin Harcourt Publishing Company

Name _____

Divide by 9

Find the quotient.

1. __4__ $= 36 \div 9$ 2. $30 \div 6 =$ ____   3. ____ $= 81 \div 9$ 4. $27 \div 9 =$ ____

5. $9 \div 9 =$ ____   6. ____ $= 63 \div 7$ 7. $36 \div 6 =$ ____   8. ____ $= 90 \div 9$

9. $9 \overline{)63}$ 10. $9 \overline{)18}$ 11. $7 \overline{)49}$ 12. $9 \overline{)45}$

Find the unknown number.

13. $48 \div 8 = g$ 14. $s = 72 \div 9$ 15. $m = 0 \div 9$ 16. $54 \div 9 = n$

    $g =$ ____ $s =$ ____ $m =$ ____ $n =$ ____

Problem Solving REAL WORLD

17. A crate of oranges has trays inside that hold 9 oranges each. There are 72 oranges in the crate. If all trays are filled, how many trays are there?

18. Van has 45 new baseball cards. He puts them in a binder that holds 9 cards on each page. How many pages does he fill?

_____   _____

© Houghton Mifflin Harcourt Publishing Company

Lesson Check

1. Darci sets up a room for a banquet. She has 54 chairs. She places 9 chairs at each table. How many tables have 9 chairs?

   (A) 5
   (B) 6
   (C) 7
   (D) 8

2. Mr. Robinson sets 36 glasses on a table. He puts the same number of glasses in each of 9 rows. How many glasses does he put in each row?

   (A) 4   (C) 6
   (B) 5   (D) 7

Spiral Review

3. Each month for 9 months, Jordan buys 2 sports books. How many more sports books does he need to buy before he has bought 25 sports books? (Lesson 4.10)

   (A) 6
   (B) 7
   (C) 8
   (D) 9

4. Find the product. (Lesson 4.8)

   $$\begin{array}{r} 8 \\ \times\, 7 \\ \hline \end{array}$$

   (A) 49
   (B) 56
   (C) 63
   (D) 64

5. Adriana made 30 pet collars to bring to the pet fair. She wants to display 3 pet collars on each hook. How many hooks will Adriana need to display all 30 pet collars? (Lesson 6.3)

   (A) 32
   (B) 12
   (C) 10
   (D) 9

6. Carla packs 4 boxes of books. Each box has 9 books. How many books does Carla pack? (Lesson 4.9)

   (A) 36
   (B) 27
   (C) 13
   (D) 5

© Houghton Mifflin Harcourt Publishing Company

Problem Solving • Two-Step Problems

Solve the problem.

1. Jack has 3 boxes of pencils with the same number of pencils in each box. His mother gives him 4 more pencils. Now Jack has 28 pencils. How many pencils are in each box?

   Think: I can start with 28 counters and act out the problem.

   _____ **8 pencils** _____

2. The art teacher has 48 paintbrushes. She puts 8 paintbrushes on each table in her classroom. How many tables are in her classroom?

   _____

3. Ricardo has 2 cases of video games with the same number of games in each case. He gives 4 games to his brother. Ricardo has 10 games left. How many video games were in each case?

   _____

4. Patty has $20 to spend on gifts for her friends. Her mother gives her $5 more. If each gift costs $5, how many gifts can she buy?

   _____

5. Joe has a collection of 35 DVD movies. He received 8 of them as gifts. Joe bought the rest of his movies over 3 years. If he bought the same number of movies each year, how many movies did Joe buy last year?

   _____

6. Liz has a 24-inch-long ribbon. She cuts nine 2-inch pieces from her original ribbon. How much of the original ribbon is left?

   _____

© Houghton Mifflin Harcourt Publishing Company

Lesson Check

1. Gavin saved $16 to buy packs of baseball cards. His father gives him $4 more. If each pack of cards costs $5, how many packs can Gavin buy?

   Ⓐ 3
   Ⓑ 4
   Ⓒ 5
   Ⓓ 6

2. Chelsea buys 8 packs of markers. Each pack contains the same number of markers. Chelsea gives 10 markers to her brother. Then, she has 54 markers left. How many markers were in each pack?

   Ⓐ 6
   Ⓑ 7
   Ⓒ 8
   Ⓓ 9

Spiral Review

3. Each foot has 5 toes. How many toes will 6 feet have? (Lesson 3.1)

   Ⓐ 11
   Ⓑ 25
   Ⓒ 30
   Ⓓ 35

4. Each month for 5 months, Sophie makes 2 quilts. How many more quilts does she need to make before she has made 16 quilts?

   (Lesson 4.10)

   Ⓐ 3
   Ⓑ 6
   Ⓒ 7
   Ⓓ 8

5. Meredith practices the piano for 3 hours each week. How many hours will she practice in 8 weeks?

   (Lesson 4.3)

   Ⓐ 18 hours
   Ⓑ 21 hours
   Ⓒ 24 hours
   Ⓓ 27 hours

6. Find the unknown factor. (Lesson 5.2)

   $9 \times \blacksquare = 36$

   Ⓐ 3
   Ⓑ 4
   Ⓒ 6
   Ⓓ 8

© Houghton Mifflin Harcourt Publishing Company

Name _____

Order of Operations

Write *correct* if the operations are listed in the correct order.
If not correct, write the correct order of operations.

1. $45 - 3 \times 5$ subtract, multiply

2. $3 \times 4 \div 2$ divide, multiply

_____**multiply, subtract**_____

_____

3. $5 + 12 \div 2$ divide, add

4. $7 \times 10 + 3$ add, multiply

_____

_____

Follow the order of operations to find the unknown number.

5. $6 + 4 \times 3 = n$

6. $8 - 3 + 2 = k$

7. $24 \div 3 + 5 = p$

$n =$ _____

$k =$ _____

$p =$ _____

8. $12 - 2 \times 5 = r$

9. $7 \times 8 - 6 = j$

10. $4 + 3 \times 9 = w$

$r =$ _____

$j =$ _____

$w =$ _____

Problem Solving REAL WORLD

11. Shelley bought 3 kites for $6 each. She gave the clerk $20. How much change should Shelley get?

12. Tim has 5 apples and 3 bags with 8 apples in each bag. How many apples does Tim have in all?

_____

_____

© Houghton Mifflin Harcourt Publishing Company

Lesson Check

1. Natalie is making doll costumes. Each costume has 4 buttons that cost 3¢ each and a zipper that costs 7¢. How much does she spend on buttons and a zipper for each costume?

Ⓐ 19¢   Ⓒ 40¢

Ⓑ 33¢   Ⓓ 49¢

2. Leonardo's mother gave him 5 bags with 6 flower bulbs in each bag to plant. He has planted all except 3 bulbs. How many flower bulbs has Leonardo planted?

Ⓐ 12   Ⓒ 27

Ⓑ 15   Ⓓ 33

Spiral Review

3. Each story in Will's apartment building is 9 feet tall. There are 10 stories in the building. How tall is the apartment building? **(Lesson 5.5)**

Ⓐ 90 feet

Ⓑ 80 feet

Ⓒ 19 feet

Ⓓ 9 feet

4. Which of the following describes a pattern in the table? **(Lesson 5.1)**

Tables	1	2	3	4
Chairs	4	8	12	16

Ⓐ Add 3.

Ⓑ Multiply by 2.

Ⓒ Subtract 3.

Ⓓ Multiply by 4.

5. For decorations, Meg cut out 8 groups of 7 snowflakes each. How many snowflakes did Meg cut out in all? **(Lesson 4.5)**

Ⓐ 72   Ⓒ 58

Ⓑ 63   Ⓓ 56

6. A small van can hold 6 students. How many small vans are needed to take 36 students on a field trip to the music museum? **(Lesson 7.6)**

Ⓐ 4   Ⓒ 7

Ⓑ 6   Ⓓ 8

© Houghton Mifflin Harcourt Publishing Company

Name _____

Chapter 7 Extra Practice

Lessons 7.1 - 7.2

Find the quotient. You may want to draw a quick picture to help.

1. $8 \div 2 =$ _____   **2.** _____ $= 14 \div 2$ **3.** $18 \div 2 =$ _____   **4.** _____ $= 12 \div 2$

5. $70 \div 10 =$ _____   **6.** $50 \div 10 =$ _____   **7.** $40 \div 10 =$ _____   **8.** $90 \div 10 =$ _____

Lessons 7.3 - 7.4

Find the quotient.

1. $15 \div 5 =$ _____   **2.** _____ $= 45 \div 5$ **3.** _____ $= 10 \div 5$ **4.** $40 \div 5 =$ _____

5. $6 \div 3 =$ _____   **6.** _____ $= 21 \div 3$ **7.** _____ $= 24 \div 3$ **8.** _____ $= 18 \div 3$

9. There are 30 balloons arranged in 6 equal groups. How many balloons are in each group?

_____

10. Mr. Song spends $27 on sports drinks. Each bottle costs $3. How many bottles does Mr. Song buy?

_____

Lesson 7.5

Find the quotient.

1. $28 \div 4 =$ _____   **2.** _____ $= 16 \div 4$ **3.** _____ $= 20 \div 4$ **4.** _____ $= 32 \div 4$

5. $4\overline{)36}$ **6.** $4\overline{)12}$ **7.** $4\overline{)24}$ **8.** $4\overline{)4}$

Find the unknown number.

9. $a = 40 \div 4$ **10.** $0 \div 4 = b$ **11.** $c = 36 \div 4$ **12.** $8 \div 4 = d$

$a =$ ____ $b =$ ____ $c =$ ____ $d =$ ____

© Houghton Mifflin Harcourt Publishing Company

Lessons 7.6 - 7.7

Find the unknown factor and quotient.

1. $7 \times \underline{\hspace{1cm}} = 35$ $35 \div 7 = \underline{\hspace{1cm}}$ 2. $6 \times \underline{\hspace{1cm}} = 54$ $54 \div 6 = \underline{\hspace{1cm}}$

3. $6 \times \underline{\hspace{1cm}} = 18$ $18 \div 6 = \underline{\hspace{1cm}}$ 4. $7 \times \underline{\hspace{1cm}} = 49$ $49 \div 7 = \underline{\hspace{1cm}}$

Find the quotient.

5. $36 \div 6 = \underline{\hspace{1cm}}$ 6. $48 \div 6 = \underline{\hspace{1cm}}$ 7. $7\overline{)63}$ 8. $7\overline{)56}$

Lessons 7.8 - 7.9

Find the quotient.

1. $40 \div 8 = \underline{\hspace{1cm}}$ 2. $\underline{\hspace{1cm}} = 24 \div 8$ 3. $72 \div 9 = \underline{\hspace{1cm}}$ 4. $\underline{\hspace{1cm}} = 81 \div 9$

Find the unknown number.

5. $36 \div 9 = m$ 6. $18 \div 9 = \blacksquare$ 7. $48 \div 8 = b$ 8. $56 \div 8 = p$

   $m = \underline{\hspace{1cm}}$ $\blacksquare = \underline{\hspace{1cm}}$ $b = \underline{\hspace{1cm}}$ $p = \underline{\hspace{1cm}}$

Lesson 7.10

1. At a store, there are 5 vases. Each vase has the same number of flowers. Sixteen flowers are sold. Now there are 24 flowers left. How many flowers were in each vase?

2. Lizzy bought 4 bags of apples. Each bag had the same number of apples. Her mom gave her 8 more apples. Now Lizzy has 36 apples. How many apples were in each bag?

_____   _____

Lesson 7.11

Follow the order of operations to find the unknown number.

1. $10 - 3 + 4 = t$ 2. $8 - 3 \times 2 = p$ 3. $24 \div 6 + 2 = w$

   $t = \underline{\hspace{1cm}}$ $p = \underline{\hspace{1cm}}$ $w = \underline{\hspace{1cm}}$

© Houghton Mifflin Harcourt Publishing Company

School-Home Letter

© Houghton Mifflin Harcourt Publishing Company

Vocabulary

denominator The part of a fraction below the line that tells how many equal parts are in the whole or in the group

equal parts Parts that are exactly the same size

fraction A number that names part of a whole or part of a group

numerator The part of a fraction above the line that tells how many equal parts are being counted

unit fraction A fraction that names 1 equal part of a whole. It has 1 as its top number, or numerator.

Dear Family,

During the next few weeks, our math class will be learning about fractions. We will learn to identify, read, and write fractions as part of a whole and as part of a group.

You can expect to see homework that provides practice with fractions.

Here is a sample of how your child will be taught to use unit fractions to find a fractional part of a group.

🔑 MODEL Find How Many in a Fractional Part of a Group

This is how we will be finding how many are in a fractional part of a group.

STEP 1

Find $\frac{1}{3}$ of 9.

Put 9 counters on your MathBoard.

STEP 2

Since you want to find $\frac{1}{3}$ of the group, there should be 3 equal groups.

STEP 3

Circle one of the groups to show $\frac{1}{3}$. Then count the number of counters in that group.

There are 3 counters in 1 group.

So, $\frac{1}{3}$ of 9 = 3.

Tips
Equal Groups or Parts

Before you name a fraction, be sure there are equal groups or parts.

Activity

Display a group of 12 objects, such as crayons. Have your child find fractional parts of the group by counting objects in equal groups. Ask your child to find these fractional groups of 12: $\frac{1}{2}$ (6), $\frac{1}{3}$ (4), $\frac{1}{4}$ (3), $\frac{1}{6}$ (2).

Carta para la casa

© Houghton Mifflin Harcourt Publishing Company

Vocabulario

denominador La parte de una fracción que está debajo de la barra y que indica cuántas partes iguales hay en el entero o en el grupo

partes iguales Las partes que son exactamente del mismo tamaño

fracción Un número que representa una parte de un todo o una parte de un grupo

numerador La parte de una fracción que está arriba de la barra y que indica cuántas partes iguales del entero se están tomando en cuenta

fracción unitaria Una fracción que se refiere a 1 parte igual de un entero. Tiene un 1 en la parte de arriba o numerador.

Querida familia,

Durante las próximas semanas, en la clase de matemáticas aprenderemos sobre las fracciones. Aprenderemos a identificar, leer y escribir fracciones como parte de un todo y como parte de un grupo.

Llevaré a la casa tareas que sirven para practicar las fracciones.

Este es un ejemplo de la manera como aprenderemos a usar fracciones para hallar una parte fraccionaria de un grupo.

🔑 MODELO Hallar cuántos hay en una parte fraccionaria de un grupo

Así es como hallaremos cuántos hay en una parte fraccionaria de un grupo.

PASO 1

Halla $\frac{1}{3}$ de 9.

Coloca 9 fichas en el *MathBoard*.

PASO 2

Como quieres hallar $\frac{1}{3}$ del grupo, debe haber 3 grupos iguales.

PASO 3

Encierra en un círculo uno de los grupos para mostrar $\frac{1}{3}$. Luego cuenta la cantidad de fichas en el grupo.

Hay 3 fichas en 1 grupo. Por lo tanto, $\frac{1}{3}$ de 9 = 3.

Pistas

Grupos o partes iguales

Antes de que nombres una fracción, asegúrate de que haya grupos o partes iguales.

Actividad

Muestre un grupo de 12 objetos, como crayolas. Pida a su hijo que halle las partes fraccionarias del grupo contando objetos en grupos iguales. Luego, pídale que halle estos grupos fraccionarios de 12: $\frac{1}{2}$ (6), $\frac{1}{3}$ (4), $\frac{1}{4}$ (3), $\frac{1}{6}$ (2).

Name _____

Equal Parts of a Whole

Write the number of equal parts.
Then write the name for the parts.

1.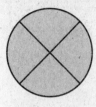

_____4_____ equal parts

___fourths___

2.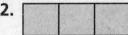

_____ equal parts

_____

3.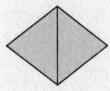

_____ equal parts

_____

4.

_____ equal parts

_____

Write whether the shape is divided into *equal* parts or *unequal* parts.

5.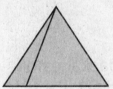

_____ parts

6.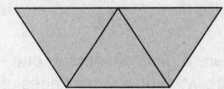

_____ parts

Problem Solving REAL WORLD

7. Diego cuts a round pizza into eight equal slices. What is the name for the parts?

_____

8. Madison is making a place mat. She divides it into 6 equal parts to color. What is the name for the parts?

_____

© Houghton Mifflin Harcourt Publishing Company

Lesson Check

1. How many equal parts are in this shape?

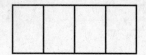

- (A) 3
- (B) 4
- (C) 5
- (D) 6

2. What is the name for the equal parts of the whole?

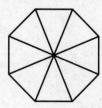

- (A) fourths
- (C) eighths
- (B) sixths
- (D) thirds

Spiral Review

3. Use a related multiplication fact to find the quotient. (Lesson 6.8)

$49 \div 7 = $ ▢

- (A) 6
- (C) 8
- (B) 7
- (D) 9

4. Find the unknown factor and quotient. (Lesson 6.8)

$9 \times $ ▢ $= 45$

$45 \div 9 = $ ▢

- (A) 4
- (C) 6
- (B) 5
- (D) 7

5. There are 5 pairs of socks in one package. Matt buys 3 packages of socks. How many pairs of socks in all does Matt buy? (Lesson 4.2)

- (A) 30
- (B) 15
- (C) 10
- (D) 8

6. Mrs. McCarr buys 9 packages of markers for an art project. Each package has 10 markers. How many markers in all does Mrs. McCarr buy? (Lesson 4.2)

- (A) 10
- (B) 19
- (C) 81
- (D) 90

© Houghton Mifflin Harcourt Publishing Company

Equal Shares

For 1–2, draw lines to show how much each person gets. Write the answer.

1. 6 friends share 3 sandwiches equally.

3 sixths of a sandwich
_____

2. 8 classmates share 4 pizzas equally.

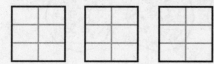

_____

3. 4 teammates share 5 granola bars equally. Draw to show how much each person gets. Shade the amount that one person gets. Write the answer.

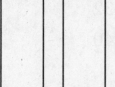    

_____

Problem Solving REAL WORLD

4. Three brothers share 2 sandwiches equally. How much of a sandwich does each brother get?

_____

_____

5. Six neighbors share 4 pies equally. How much of a pie does each neighbor get?

_____

_____

© Houghton Mifflin Harcourt Publishing Company

Lesson Check

1. Two friends share 3 fruit bars equally. How much does each friend get?

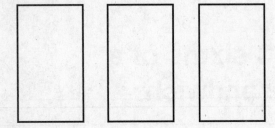

- (A) 1 half
- (B) 2 thirds
- (C) 2 halves
- (D) 3 halves

2. Four brothers share 3 pizzas equally. How much of a pizza does each brother get?

- (A) 3 halves
- (B) 4 thirds
- (C) 3 fourths
- (D) 2 fourths

Spiral Review

3. Find the quotient. (Lesson 7.4)

$3\overline{)27}$

- (A) 6
- (B) 7
- (C) 8
- (D) 9

4. Tyrice put 4 cookies in each of 7 bags. How many cookies in all did he put in the bags? (Lesson 4.5)

- (A) 11
- (B) 28
- (C) 32
- (D) 40

5. Ryan earns $5 per hour raking leaves. He earned $35. How many hours did he rake leaves? (Lesson 7.3)

- (A) 5 hours
- (B) 6 hours
- (C) 7 hours
- (D) 35 hours

6. Hannah has 229 horse stickers and 164 kitten stickers. How many more horse stickers than kitten stickers does Hannah have? (Lesson 1.10)

- (A) 45
- (B) 65
- (C) 145
- (D) 293

© Houghton Mifflin Harcourt Publishing Company

Name _____

Unit Fractions of a Whole

Write the number of equal parts in the whole.
Then write the fraction that names the shaded part.

1.

_____**6**_____ equal parts

_____$\frac{1}{6}$_____

2.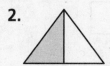

_____ equal parts

_____

3.

_____ equal parts

_____

4.

_____ equal parts

_____

Draw a picture of the whole.

5. $\frac{1}{3}$ is

6. $\frac{1}{8}$ is

Problem Solving REAL WORLD

7. Tyler made a pan of cornbread. He cut it into 8 equal pieces and ate 1 piece. What fraction of the cornbread did Tyler eat?

_____

8. Anna cut an apple into 4 equal pieces. She gave 1 piece to her sister. What fraction of the apple did Anna give to her sister?

_____

© Houghton Mifflin Harcourt Publishing Company

Lesson Check

1. What fraction names the shaded part?

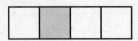

(A) $\frac{1}{3}$

(B) $\frac{1}{4}$

(C) $\frac{1}{6}$

(D) $\frac{1}{8}$

2. Tasha cut a fruit bar into 3 equal parts. She ate 1 part. What fraction of the fruit bar did Tasha eat?

(A) $\frac{1}{2}$

(B) $\frac{1}{3}$

(C) $\frac{1}{4}$

(D) $\frac{1}{6}$

Spiral Review

3. Alex has 5 lizards. He divides them equally among 5 cages. How many lizards does Alex put in each cage? (Lesson 6.9)

(A) 0

(B) 1

(C) 5

(D) 10

4. Find the product. (Lesson 3.7)

$8 \times 1 = $

(A) 0

(B) 1

(C) 8

(D) 9

5. Leo bought 6 chew toys for his new puppy. Each chew toy cost $4. How much did Leo spend in all for the chew toys? (Lesson 4.1)

(A) $10

(B) $12

(C) $18

(D) $24

6. Lilly is making a picture graph. Each picture of a star is equal to two books she has read. The row for the month of December has 3 stars. How many books did Lilly read during the month of December? (Lesson 2.2)

(A) 3

(B) 5

(C) 6

(D) 9

© Houghton Mifflin Harcourt Publishing Company

Fractions of a Whole

Write the fraction that names each part. Write a fraction
in words and in numbers to name the shaded part.

1. 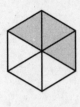 Each part is ___ $\frac{1}{6}$ ___.

_____three_____ sixths

_____ $\frac{3}{6}$ _____

2. 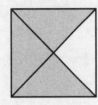 Each part is _____.

_____ eighths

_____

3.   Each part is _____.

_____ thirds

_____

4.   Each part is _____.

_____ fourths

_____

Shade the fraction circle to model the fraction.
Then write the fraction in numbers.

5. four out of six

_____

6. eight out of eight

_____

Problem Solving REAL WORLD

7. Emma makes a poster for the
school's spring concert. She divides
the poster into 8 equal parts. She
uses two of the parts for the title.
What fraction of the poster does
Emma use for the title?

8. Lucas makes a flag. It has 6 equal
parts. Five of the parts are red. What
fraction of the flag is red?

_____

© Houghton Mifflin Harcourt Publishing Company

Lesson Check

1. What fraction names the shaded part?

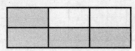

- Ⓐ $\frac{4}{6}$
- Ⓒ $\frac{4}{8}$
- Ⓑ $\frac{2}{4}$
- Ⓓ $\frac{2}{6}$

2. What fraction names the shaded part?

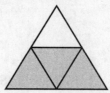

- Ⓐ one fourth
- Ⓑ one third
- Ⓒ three fourths
- Ⓓ four thirds

Spiral Review

3. Sarah biked for 115 minutes last week. Jennie biked for 89 minutes last week. How many minutes in all did the girls bike? **(Lesson 1.7)**

- Ⓐ 26 minutes
- Ⓒ 204 minutes
- Ⓑ 194 minutes
- Ⓓ 294 minutes

4. Harrison made a building using 124 blocks. Greyson made a building using 78 blocks. How many more blocks did Harrison use than Greyson did? **(Lesson 1.10)**

- Ⓐ 46
- Ⓒ 154
- Ⓑ 56
- Ⓓ 202

5. Von bought a bag of 24 dog treats. He gives his puppy 3 treats a day. How many days will the bag of dog treats last? **(Lesson 7.4)**

- Ⓐ 3 days
- Ⓑ 6 days
- Ⓒ 8 days
- Ⓓ 21 days

6. How many students chose swimming? **(Lesson 2.2)**

Favorite Activity	
Skating	☺ ☺
Swimming	☺ ☺ ☺ ☺ ☺
Biking	☺ ☺ ☺ ☺
Key: Each ☺ = 5 votes.	

- Ⓐ 5
- Ⓒ 20
- Ⓑ 10
- Ⓓ 25

© Houghton Mifflin Harcourt Publishing Company

Fractions on a Number Line

Use fraction strips to help you complete the number line. Then locate and draw a point for the fraction.

1. $\frac{1}{3}$

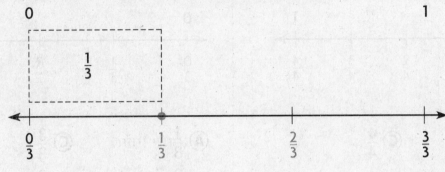

2. $\frac{3}{4}$

Write the fraction that names the point.

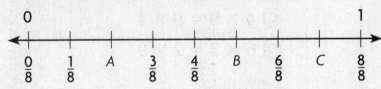

3. point A _____

4. point B _____

5. point C _____

Problem Solving REAL WORLD

6. Jade ran 6 times around her neighborhood to complete a total of 1 mile. How many times will she need to run to complete $\frac{5}{6}$ of a mile?

_____

7. A missing fraction on a number line is located exactly halfway between $\frac{3}{6}$ and $\frac{5}{6}$. What is the missing fraction?

_____

© Houghton Mifflin Harcourt Publishing Company

Lesson Check

1. Which fraction names point G on the number line?

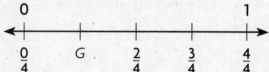

(A) $\frac{1}{4}$ (C) $\frac{4}{4}$

(B) $\frac{2}{4}$ (D) $\frac{4}{1}$

2. Which fraction names point R on the number line?

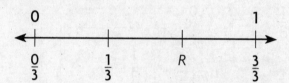

(A) $\frac{1}{3}$ (C) $\frac{3}{3}$

(B) $\frac{2}{3}$ (D) $\frac{3}{2}$

Spiral Review

3. Each table in the cafeteria can seat 10 students. How many tables are needed to seat 40 students?

(Lesson 7.2)

(A) 10   (C) 5

(B) 8   (D) 4

4. Which is an example of the Commutative Property of Multiplication? (Lesson 3.6)

(A) $6 \times 1 = 6 \times 1$

(B) $4 + 9 = 4 \times 9$

(C) $4 \times 9 = 9 \times 4$

(D) $6 \times 3 = 2 \times 9$

5. Pedro shaded part of a circle. Which fraction names the shaded part? (Lesson 8.4)

(A) $\frac{1}{8}$ (C) $\frac{7}{8}$

(B) $\frac{1}{7}$ (D) $\frac{8}{7}$

6. Which is true?

(Lesson 6.9)

(A) $8 \div 1 = 8$

(B) $8 \div 8 = 8$

(C) $8 \times 0 = 8$

(D) $1 = 8 \times 1$

© Houghton Mifflin Harcourt Publishing Company

Name _____

Relate Fractions and Whole Numbers

Use the number line to find whether the two numbers
are equal. Write *equal* or *not equal*.

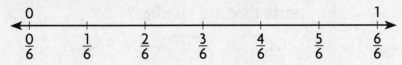

1. $\frac{0}{6}$ and 1

2. 1 and $\frac{6}{6}$

3. $\frac{1}{6}$ and $\frac{6}{6}$

__not equal__   _____   _____

Each shape is **1** whole. Write a fraction greater than **1**
for the parts that are shaded.

4. 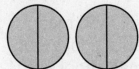   2 = _____

5. 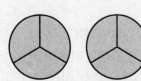   4 = _____

6.    3 = _____

7.   1 = _____

Problem Solving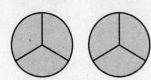

8. Rachel jogged along a trail that was
$\frac{1}{4}$ of a mile long. She jogged along
the trail 8 times. How many miles
did Rachel jog in all?

9. Jon ran around a track that was $\frac{1}{8}$ of
a mile long. He ran around the track
24 times. How many miles did Jon
run in all?

_____   _____

© Houghton Mifflin Harcourt Publishing Company

Lesson Check

1. Each shape is 1 whole. Which fraction greater than 1 names the parts that are shaded?

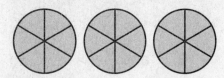

Ⓐ $\frac{6}{18}$ Ⓒ $\frac{6}{3}$

Ⓑ $\frac{3}{6}$ Ⓓ $\frac{18}{6}$

2. Each shape is 1 whole. Which fraction greater than 1 names the parts that are shaded?

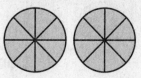

Ⓐ $\frac{8}{2}$ Ⓒ $\frac{8}{16}$

Ⓑ $\frac{16}{8}$ Ⓓ $\frac{2}{8}$

Spiral Review

3. Tara has 598 pennies and 231 nickels. How many pennies and nickels does she have in all? (Lesson 1.7)

$$598$$
$$+\ 231$$

Ⓐ 719   Ⓒ 819

Ⓑ 729   Ⓓ 829

4. Dylan read 6 books. Kylie read double the number of books that Dylan read. How many books did Kylie read? (Lesson 4.1)

Ⓐ 4

Ⓑ 8

Ⓒ 12

Ⓓ 14

5. Alyssa divides a granola bar into halves. How many equal parts are there? (Lesson 8.1)

Ⓐ 2   Ⓒ 4

Ⓑ 3   Ⓓ 6

6. There are 4 students in each small reading group. If there are 24 students in all, how many reading groups are there? (Lesson 7.5)

Ⓐ 5   Ⓒ 7

Ⓑ 6   Ⓓ 8

© Houghton Mifflin Harcourt Publishing Company

Name _____

Fractions of a Group

Write a fraction to name the shaded part of each group.

1.

$\dfrac{6}{8}$ _____

2. _____

Write a whole number and a fraction greater
than 1 to name the part filled. Think: 1 container = 1

3.

4.

_____ _____   _____

Draw a quick picture. Then, write a fraction
to name the shaded part of the group.

5. Draw 4 circles.
   Shade 2 circles.

6. Draw 6 circles.
   Make 3 groups.
   Shade 1 group.

_____   _____

Problem Solving

7. Brian has 3 basketball cards and
   5 baseball cards. What fraction of
   Brian's cards are baseball cards?

8. Sophia has 3 pink tulips and 3 white
   tulips. What fraction of Sophia's tulips
   are pink?

_____   _____

© Houghton Mifflin Harcourt Publishing Company

Lesson Check

1. What fraction of the group is shaded?

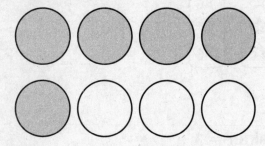

Ⓐ $\frac{5}{3}$ Ⓒ $\frac{3}{5}$

Ⓑ $\frac{5}{8}$ Ⓓ $\frac{3}{8}$

2. What fraction of the group is shaded?

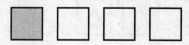

Ⓐ $\frac{1}{4}$

Ⓑ $\frac{1}{2}$

Ⓒ $\frac{2}{4}$

Ⓓ $\frac{4}{1}$

Spiral Review

3. Which number sentence does the array represent? (Lesson 4.5)

Ⓐ $4 \times 7 = 28$

Ⓑ $3 \times 8 = 24$

Ⓒ $3 \times 7 = 21$

Ⓓ $3 \times 6 = 18$

4. Juan has 436 baseball cards and 189 football cards. How many more baseball cards than football cards does Juan have? (Lesson 1.10)

Ⓐ 625

Ⓑ 353

Ⓒ 347

Ⓓ 247

5. Sydney bought 3 bottles of glitter. Each bottle of glitter cost $6. How much did Sydney spend in all on the bottles of glitter? (Lesson 4.3)

Ⓐ $24   Ⓒ $12

Ⓑ $18   Ⓓ $9

6. Add. (Lesson 1.7)

$$\begin{array}{r} 262 \\ + 119 \\ \hline \end{array}$$

Ⓐ 143   Ⓒ 381

Ⓑ 371   Ⓓ 481

© Houghton Mifflin Harcourt Publishing Company

Find Part of a Group
Using Unit Fractions

**Circle equal groups to solve. Count the number of
items in 1 group.**

1. $\frac{1}{4}$ of 12 = **3**

2. $\frac{1}{8}$ of 16 = _____

3. $\frac{1}{3}$ of 12 = _____

4. $\frac{1}{3}$ of 9 = _____

5. $\frac{1}{6}$ of 18 = _____

6. $\frac{1}{2}$ of 4 = _____

Problem Solving REAL WORLD

7. Marco drew 24 pictures. He drew
$\frac{1}{6}$ of them in art class. How many
pictures did Marco draw in art class?

8. Caroline has 16 marbles. One eighth
of them are blue. How many of
Caroline's marbles are blue?

_____

_____

© Houghton Mifflin Harcourt Publishing Company

Lesson Check

1. Ms. Davis made 12 blankets for her grandchildren. One third of the blankets are blue. How many blue blankets did she make?

○ ○ ○
○ ○ ○
○ ○ ○
○ ○ ○

Ⓐ 3   Ⓒ 9
Ⓑ 4   Ⓓ 12

2. Jackson mowed 16 lawns. One fourth of the lawns are on Main Street. How many lawns on Main Street did Jackson mow?

○ ○ ○ ○
○ ○ ○ ○
○ ○ ○ ○
○ ○ ○ ○

Ⓐ 4   Ⓒ 8
Ⓑ 6   Ⓓ 12

Spiral Review

3. Find the difference. (Lesson 1.10)

$$509$$
$$-175$$

Ⓐ 334
Ⓑ 374
Ⓒ 434
Ⓓ 474

4. Find the quotient. (Lesson 7.6)

$$6)\overline{54}$$

Ⓐ 6
Ⓑ 7
Ⓒ 8
Ⓓ 9

5. There are 226 pets entered in the pet show. What is 226 rounded to the nearest hundred? (Lesson 1.2)

Ⓐ 200   Ⓒ 300
Ⓑ 220   Ⓓ 400

6. Ladonne made 36 muffins. She put the same number of muffins on each of 4 plates. How many muffins did she put on each plate? (Lesson 7.5)

Ⓐ 3   Ⓒ 9
Ⓑ 6   Ⓓ 12

© Houghton Mifflin Harcourt Publishing Company

Problem Solving • Find the Whole Group Using Unit Fractions

Draw a quick picture to solve.

1. Katrina has 2 blue ribbons for her hair. One fourth of all her ribbons are blue. How many ribbons does Katrina have in all?

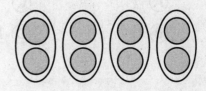

8 ribbons

2. One eighth of Tony's books are mystery books. He has 3 mystery books. How many books does Tony have in all?

_____

3. Brianna has 4 pink bracelets. One third of all her bracelets are pink. How many bracelets does Brianna have?

_____

4. Ramal filled 3 pages in a stamp album. This is one sixth of the pages in the album. How many pages are there in Ramal's stamp album?

_____

5. Jeff helped repair one half of the bicycles in a bike shop last week. If Jeff worked on 5 bicycles, how many bicycles did the shop repair in all last week?

_____

6. Layla collects postcards. She has 7 postcards from Europe. Her postcards from Europe are one third of her total collection. How many postcards in all does Layla have?

_____

© Houghton Mifflin Harcourt Publishing Company

Lesson Check

1. A zoo has 2 male lions. One sixth of the lions are male lions. How many lions are there at the zoo?

   Ⓐ 4 Ⓒ 8

   Ⓑ 6 Ⓓ 12

2. Max has 5 red model cars. One third of his model cars are red. How many model cars does Max have?

   Ⓐ 15 Ⓒ 10

   Ⓑ 12 Ⓓ 8

Spiral Review

3. There are 382 trees in the local park. What is the number of trees rounded to the nearest hundred? (Lesson 1.2)

   Ⓐ 300

   Ⓑ 380

   Ⓒ 400

   Ⓓ 500

4. The Jones family is driving 458 miles on their vacation. So far, they have driven 267 miles. How many miles do they have left to drive? (Lesson 1.10)

   $$\begin{array}{r} 458 \\ -\ 267 \end{array}$$

   Ⓐ 191 miles Ⓒ 211 miles

   Ⓑ 201 miles Ⓓ 291 miles

5. Ken has 6 different colors of marbles. He has 9 marbles of each color. How many marbles does Ken have in all? (Lesson 4.3)

   Ⓐ 15

   Ⓑ 45

   Ⓒ 54

   Ⓓ 63

6. Eight friends share two pizzas equally. How much of a pizza does each friend get? (Lesson 8.2)

   Ⓐ 8 halves Ⓒ 2 sixths

   Ⓑ 4 eighths Ⓓ 2 eighths

© Houghton Mifflin Harcourt Publishing Company

Name _____

Chapter 8 Extra Practice

Lesson 8.1

Write the number of equal parts. Then write the name for the parts.

1.

_____ equal parts

_____

2.

_____ equal parts

_____

3.

_____ equal parts

_____

Lesson 8.2

Draw lines to show how much each person gets. Write the answer.

1. 4 friends share 3 oranges equally.

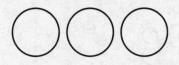

_____

_____

2. 6 sisters share 4 sandwiches equally.

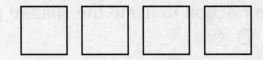

_____

_____

Lessons 8.3 - 8.4

Write the number of equal parts in the whole. Write a fraction in words and in numbers to name the shaded part.

1.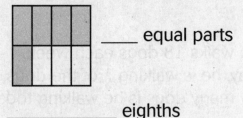

_____ equal parts

_____ eighths

_____

2.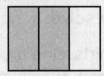

_____ equal parts

_____ thirds

_____

© Houghton Mifflin Harcourt Publishing Company

Lesson 8.5

Write the fraction that names the point.

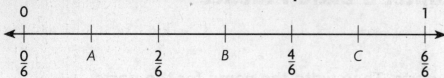

1. point A ____   2. point B ____   3. point C ____

Lesson 8.6

Each shape is 1 whole. Write a fraction
greater than 1 for the parts that are shaded.

1.
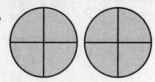

2 = ____

2.
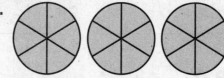

3 = ____

Lesson 8.7

Write a fraction to name the shaded part of each group.

1.
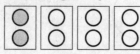

2.

_____   _____

Lessons 8.8 - 8.9

Draw a quick picture to solve.

1. Charlotte has 12 T-shirts. One fourth
   of her T-shirts are green. How many
   of Charlotte's T-shirts are green?

2. Josh walks 18 dogs each week.
   Today, he is walking $\frac{1}{3}$ of the dogs.
   How many dogs is he walking today?

_____   _____

© Houghton Mifflin Harcourt Publishing Company

School-Home Letter

© Houghton Mifflin Harcourt Publishing Company

Dear Family,

During the next few weeks, our math class will be learning more about fractions. We will learn how to compare fractions, order fractions, and find equivalent fractions.

You can expect to see homework that provides practice with fractions.

Here is a sample of how your child will be taught to compare fractions that have the same numerator.

Vocabulary

equivalent fractions Two or more fractions that name the same amount

greater than (>) A symbol used to compare two numbers, with the greater number given first

less than (<) A symbol used to compare two numbers, with the lesser number given first

🔑 MODEL Compare Fractions with the Same Numerator

This is one way we will be comparing fractions that have the same numerator.

STEP 1

Compare $\frac{4}{8}$ and $\frac{4}{6}$.

Look at the numerators.

Each numerator is 4.

The numerators are the same.

STEP 2

Since the numerators are the same, look at the denominators, 8 and 6.

The more pieces a whole is divided into, the smaller the pieces are. Eighths are smaller pieces than sixths.

So, $\frac{4}{8}$ is a smaller fraction of the whole than $\frac{4}{6}$.

$\frac{4}{8}$ is less than $\frac{4}{6}$. $\frac{4}{8} < \frac{4}{6}$

Tips

Identifying Fewer Pieces

The fewer pieces a whole is divided into, the larger the pieces are. For example, when a whole is divided into 6 equal pieces, the pieces are larger than when the same size whole is divided into 8 equal pieces. So, $\frac{4}{6}$ is greater than (>) $\frac{4}{8}$.

Activity

Play a card game to help your child practice comparing fractions. On several cards, write a pair of fractions with the same numerator and draw a circle between the fractions. Players take turns drawing a card and telling whether *greater than* (>) or *less than* (<) belongs in the circle.

Capítulo 9

Carta para la casa

Vocabulario

fracciones equivalentes Dos o más fracciones que representan la misma cantidad

mayor que Símbolo que se usa para comparar dos números. El número mayor se escribe primero (>).

menor que Símbolo que se usa para comparar dos números. El número menor se escribe primero (<).

Querida familia,

Durante las próximas semanas, en la clase de matemáticas aprenderemos más sobre las fracciones. Aprenderemos a comparar y ordenar fracciones, y a hallar fracciones equivalentes.

Llevaré a la casa tareas para practicar las fracciones.

Este es un ejemplo de la manera como aprenderemos a comparar fracciones que tienen el mismo numerador.

🔑 MODELO Comparar fracciones que tienen el mismo denominador

Esta es una manera como compararemos fracciones que tienen el mismo numerador.

Paso 1

Compara $\frac{4}{8}$ y $\frac{4}{6}$.

Mira los numeradores.

Cada numerador es 4.

Los numeradores son iguales.

Paso 2

Dado que los numeradores son iguales, Mira los denominadores 8 y 6.

Entre más piezas se divida un entero, las piezas serán más pequeñas. Los octavos son piezas más pequeñas que los sextos.

Por lo tanto, $\frac{4}{8}$ es una fracción menor del entero que $\frac{4}{6}$.

$\frac{4}{8}$ es menor que $\frac{4}{6}$. $\frac{4}{8} < \frac{4}{6}$

Pistas

Identificar menos piezas

Entre menos piezas se divida un entero, las piezas serán más grandes. Por ejemplo, si un entero se divide en 6 piezas iguales, las piezas son más grandes que las piezas del mismo entero, si éste se divide en 8 piezas iguales. Por lo tanto, $\frac{4}{6}$ es mayor que (>) $\frac{4}{8}$.

Actividad

Ayude a su hijo a comparar fracciones jugando con tarjetas de fracciones. En varias tarjetas, escriba pares de fracciones con el mismo numerador y dibuje un círculo entre las fracciones. Túrnense para dibujar cada tarjeta y decir qué debe ir en el círculo: "mayor que" o "menor que."

© Houghton Mifflin Harcourt Publishing Company

Problem Solving • Compare Fractions

Solve.

1. Luis skates $\frac{2}{3}$ mile from his home to school. Isabella skates $\frac{2}{4}$ mile to get to school. Who skates farther?

   Think: Use fraction strips to act it out.

   Luis

2. Sandra makes a pizza. She puts mushrooms on $\frac{2}{8}$ of the pizza. She adds green peppers to $\frac{5}{8}$ of the pizza. Which topping covers more of the pizza?

3. The jars of paint in the art room have different amounts of paint. The green paint jar is $\frac{4}{8}$ full. The purple paint jar is $\frac{4}{6}$ full. Which paint jar is less full?

4. Jan has a recipe for bread. She uses $\frac{2}{3}$ cup of flour and $\frac{1}{3}$ cup of chopped onion. Which ingredient does she use more of, flour or onion?

5. Edward walked $\frac{3}{4}$ mile from his home to the park. Then he walked $\frac{2}{4}$ mile from the park to the library. Which distance is shorter?

© Houghton Mifflin Harcourt Publishing Company

Lesson Check

1. Ali and Jonah collect seashells in identical buckets. When they are finished, Ali's bucket is $\frac{2}{6}$ full and Jonah's bucket is $\frac{3}{6}$ full. Which of the following correctly compares the fractions?

   Ⓐ $\frac{2}{6} = \frac{3}{6}$ Ⓒ $\frac{3}{6} < \frac{2}{6}$

   Ⓑ $\frac{2}{6} > \frac{3}{6}$ Ⓓ $\frac{3}{6} > \frac{2}{6}$

2. Rosa paints a wall in her bedroom. She puts green paint on $\frac{5}{8}$ of the wall and blue paint on $\frac{3}{8}$ of the wall. Which of the following correctly compares the fractions?

   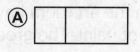

   Ⓐ $\frac{5}{8} > \frac{3}{8}$ Ⓒ $\frac{3}{8} > \frac{5}{8}$

   Ⓑ $\frac{5}{8} < \frac{3}{8}$ Ⓓ $\frac{3}{8} = \frac{5}{8}$

Spiral Review

3. Dan divides a pie into eighths. How many equal parts are there? (Lesson 8.1)

   Ⓐ 3

   Ⓑ 6

   Ⓒ 8

   Ⓓ 10

4. Which shows equal parts? (Lesson 8.1)

5. Charles places 30 pictures on his bulletin board in 6 equal rows. How many pictures are in each row?

   (Lesson 6.7)

   Ⓐ 3

   Ⓑ 4

   Ⓒ 5

   Ⓓ 6

6. Which of the following describes a pattern in the table? (Lesson 5.1)

   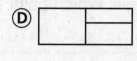

Tables	1	2	3	4	5
Chairs	5	10	15	20	25

   Ⓐ Add 1.   Ⓒ Multiply by 2.

   Ⓑ Add 4.   Ⓓ Multiply by 5.

© Houghton Mifflin Harcourt Publishing Company

Name _____

Compare Fractions with the Same Denominator

Compare. Write <, >, or =.

1. $\frac{3}{4}$ ⊝ $\frac{1}{4}$ (>)

2. $\frac{3}{6}$ ◯ $\frac{0}{6}$

3. $\frac{1}{2}$ ◯ $\frac{1}{2}$

4. $\frac{5}{6}$ ◯ $\frac{6}{6}$

5. $\frac{7}{8}$ ◯ $\frac{5}{8}$

6. $\frac{2}{3}$ ◯ $\frac{3}{3}$

7. $\frac{8}{8}$ ◯ $\frac{0}{8}$

8. $\frac{1}{6}$ ◯ $\frac{1}{6}$

9. $\frac{3}{4}$ ◯ $\frac{2}{4}$

10. $\frac{1}{6}$ ◯ $\frac{2}{6}$

11. $\frac{1}{2}$ ◯ $\frac{0}{2}$

12. $\frac{3}{8}$ ◯ $\frac{3}{8}$

13. $\frac{1}{4}$ ◯ $\frac{4}{4}$

14. $\frac{5}{8}$ ◯ $\frac{4}{8}$

15. $\frac{4}{6}$ ◯ $\frac{6}{6}$

Problem Solving  REAL WORLD

16. Ben mowed $\frac{5}{6}$ of his lawn in one hour. John mowed $\frac{4}{6}$ of his lawn in one hour. Who mowed less of his lawn in one hour?

17. Darcy baked 8 muffins. She put blueberries in $\frac{5}{8}$ of the muffins. She put raspberries in $\frac{3}{8}$ of the muffins. Did more muffins have blueberries or raspberries?

_____

_____

© Houghton Mifflin Harcourt Publishing Company

Lesson Check

1. Julia paints $\frac{2}{6}$ of a wall in her room white. She paints more of the wall green. Which fraction could show the part of the wall that is green?

   Ⓐ $\frac{1}{6}$ Ⓒ $\frac{3}{6}$

   Ⓑ $\frac{2}{6}$ Ⓓ $\frac{0}{6}$

2. Liam is comparing fraction circles. Which of the following statements is true?

   Ⓐ $\frac{1}{2} = \frac{1}{2}$ Ⓒ $\frac{4}{6} < \frac{3}{6}$

   Ⓑ $\frac{3}{4} > \frac{4}{4}$ Ⓓ $\frac{2}{8} = \frac{3}{8}$

Spiral Review

3. Mr. Edwards buys 2 new knobs for each of his kitchen cabinets. The kitchen has 9 cabinets. How many knobs does he buy? (Lesson 4.1)

   Ⓐ 20

   Ⓑ 18

   Ⓒ 16

   Ⓓ 12

4. Allie builds a new bookcase with 8 shelves. She can put 30 books on each shelf. How many books can the bookcase hold? (Lesson 5.4)

   Ⓐ 30

   Ⓑ 38

   Ⓒ 240

   Ⓓ 300

5. The Good Morning Café has 28 customers for breakfast. There are 4 people sitting at each table. How many tables are filled?

   (Lesson 7.5)

   Ⓐ 8

   Ⓑ 7

   Ⓒ 6

   Ⓓ 4

6. Ella wants to use the Commutative Property of Multiplication to help find the product 5×4. Which number sentence can she use?

   (Lesson 3.6)

   Ⓐ $5 + 4 = 9$

   Ⓑ $5 \times 5 = 25$

   Ⓒ $5 - 4 = 1$

   Ⓓ $4 \times 5 = 20$

© Houghton Mifflin Harcourt Publishing Company

Compare Fractions with the Same Numerator

Compare. Write <, >, or =.

1. $\frac{1}{8}$ $<$ $\frac{1}{2}$

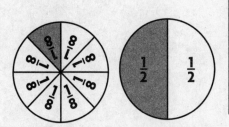

2. $\frac{3}{8}$ ◯ $\frac{3}{6}$

3. $\frac{2}{3}$ ◯ $\frac{2}{4}$

4. $\frac{2}{8}$ ◯ $\frac{2}{3}$

5. $\frac{3}{6}$ ◯ $\frac{3}{4}$

6. $\frac{1}{2}$ ◯ $\frac{1}{6}$

7. $\frac{5}{6}$ ◯ $\frac{5}{8}$

8. $\frac{4}{8}$ ◯ $\frac{4}{8}$

9. $\frac{6}{8}$ ◯ $\frac{6}{6}$

Problem Solving

10. Javier is buying food in the lunch line. The tray of salad plates is $\frac{3}{8}$ full. The tray of fruit plates is $\frac{3}{4}$ full. Which tray is more full?

_____

11. Rachel bought some buttons. Of the buttons, $\frac{2}{4}$ are yellow and $\frac{2}{8}$ are red. Rachel bought more of which color buttons?

_____

© Houghton Mifflin Harcourt Publishing Company

Lesson Check

1. Which symbol makes the statement true?

   $\frac{3}{4}$ ● $\frac{3}{8}$

   Ⓐ >
   Ⓑ <
   Ⓒ =
   Ⓓ none

2. Which symbol makes the statement true?

   $\frac{2}{4}$ ● $\frac{2}{3}$

   Ⓐ >
   Ⓑ <
   Ⓒ =
   Ⓓ none

Spiral Review

3. Anita divided a circle into 6 equal parts and shaded 1 of the parts. Which fraction names the part she shaded? (Lesson 8.3)

   Ⓐ $\frac{1}{6}$ Ⓒ $\frac{5}{6}$

   Ⓑ $\frac{1}{5}$ Ⓓ $\frac{1}{1}$

4. Which fraction names the shaded part of the rectangle? (Lesson 8.4)

   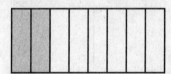

   Ⓐ $\frac{1}{8}$ Ⓒ $\frac{6}{8}$

   Ⓑ $\frac{2}{8}$ Ⓓ $\frac{8}{8}$

5. Chip worked at the animal shelter for 6 hours each week for several weeks. He worked for a total of 42 hours. Which of the following can be used to find the number of weeks Chip worked at the animal shelter? (Lesson 7.6)

   Ⓐ 6 + 42

   Ⓑ 42 − 6

   Ⓒ 42 ÷ 6

   Ⓓ 42 × 6

6. Mr. Jackson has 20 quarters. If he gives 4 quarters to each of his children, how many children does Mr. Jackson have? (Lesson 7.5)

   Ⓐ 3

   Ⓑ 4

   Ⓒ 5

   Ⓓ 6

© Houghton Mifflin Harcourt Publishing Company

Name _____

Compare Fractions

Compare. Write <, >, or =. Write the strategy you used.

1. $\frac{3}{8}$ ⊘ $\frac{3}{4}$ <

   Think: The numerators are the same. Compare the denominators. The greater fraction will have the lesser denominator.

   same numerator _____

   _____

2. $\frac{2}{3}$ ◯ $\frac{7}{8}$

   _____

   _____

3. $\frac{3}{4}$ ◯ $\frac{1}{4}$

   _____

   _____

Name a fraction that is less than or greater than the given fraction. Draw to justify your answer.

4. greater than $\frac{1}{3}$ —

5. less than $\frac{3}{4}$ —

Problem Solving REAL WORLD

6. At the third-grade party, two groups each had their own pizza. The blue group ate $\frac{7}{8}$ pizza. The green group ate $\frac{2}{8}$ pizza. Which group ate more of their pizza?

   _____

7. Ben and Antonio both take the same bus to school. Ben's ride is $\frac{7}{8}$ mile. Antonio's ride is $\frac{3}{4}$ mile. Who has a longer bus ride?

   _____

© Houghton Mifflin Harcourt Publishing Company

Lesson Check

1. Which statement is correct?

   (A) $\frac{2}{3} > \frac{7}{8}$

   (B) $\frac{2}{3} < \frac{7}{8}$

   (C) $\frac{2}{3} = \frac{7}{8}$

   (D) $\frac{7}{8} < \frac{2}{3}$

2. Which symbol makes the statement true?

   $\frac{2}{4}$ ● $\frac{2}{6}$

   (A) >

   (B) <

   (C) =

   (D) none

Spiral Review

3. Cam, Stella, and Rose each picked 40 apples. They put all their apples in one crate. How many apples are in the crate? **(Lesson 5.5)**

   (A) 40

   (B) 43

   (C) 120

   (D) 123

4. Each shape is 1 whole. Which fraction is represented by the shaded part of the model? **(Lesson 8.6)**

   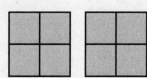

   (A) $\frac{2}{4}$ (C) $\frac{8}{4}$

   (B) $\frac{4}{4}$ (D) $\frac{8}{1}$

5. Which related multiplication fact can you use to find $16 \div \blacksquare = 2$? **(Lesson 7.8)**

   (A) $4 \times 4 = 16$

   (B) $8 \times 2 = 16$

   (C) $8 \times 1 = 8$

   (D) $4 \times 2 = 8$

6. What is the unknown factor? **(Lesson 5.2)**

   $9 \times \blacksquare = 36$

   (A) 7

   (B) 6

   (C) 4

   (D) 3

© Houghton Mifflin Harcourt Publishing Company

Compare and Order Fractions

Write the fractions in order from greatest to least.

1. $\frac{4}{4}, \frac{1}{4}, \frac{3}{4}$ $\dfrac{4}{4}$, $\dfrac{3}{4}$, $\dfrac{1}{4}$

   Think: The denominators are the same, so compare the numerators: $4 > 3 > 1$.

2. $\frac{2}{8}, \frac{5}{8}, \frac{1}{8}$ _____, _____, _____

3. $\frac{1}{3}, \frac{1}{6}, \frac{1}{2}$ _____, _____, _____

4. $\frac{2}{3}, \frac{2}{6}, \frac{2}{8}$ _____, _____, _____

Write the fractions in order from least to greatest.

5. $\frac{2}{4}, \frac{4}{4}, \frac{3}{4}$ _____, _____, _____

6. $\frac{4}{6}, \frac{5}{6}, \frac{2}{6}$ _____, _____, _____

7. $\frac{7}{8}, \frac{0}{8}, \frac{3}{8}$ _____, _____, _____

8. $\frac{3}{4}, \frac{3}{6}, \frac{3}{8}$ _____, _____, _____

Problem Solving REAL WORLD

9. Mr. Jackson ran $\frac{7}{8}$ mile on Monday. He ran $\frac{3}{8}$ mile on Wednesday and $\frac{5}{8}$ mile on Friday. On which day did Mr. Jackson run the shortest distance?

10. Delia has three pieces of ribbon. Her red ribbon is $\frac{2}{4}$ foot long. Her green ribbon is $\frac{2}{3}$ foot long. Her yellow ribbon is $\frac{2}{6}$ foot long. She wants to use the longest piece for a project. Which color ribbon should Delia use?

_____   _____

© Houghton Mifflin Harcourt Publishing Company

Lesson Check

1. Which list orders the fractions from least to greatest?

   Ⓐ $\frac{1}{8}, \frac{1}{3}, \frac{1}{6}$

   Ⓑ $\frac{1}{3}, \frac{1}{6}, \frac{1}{8}$

   Ⓒ $\frac{1}{8}, \frac{1}{6}, \frac{1}{3}$

   Ⓓ $\frac{1}{6}, \frac{1}{8}, \frac{1}{3}$

2. Which list orders the fractions from greatest to least?

   Ⓐ $\frac{3}{8}, \frac{3}{6}, \frac{3}{4}$

   Ⓑ $\frac{3}{4}, \frac{3}{6}, \frac{3}{8}$

   Ⓒ $\frac{3}{4}, \frac{3}{8}, \frac{3}{4}$

   Ⓓ $\frac{3}{6}, \frac{3}{4}, \frac{3}{8}$

Spiral Review

3. What fraction of the group of cars is shaded? (Lesson 8.7)

   Ⓐ $\frac{3}{8}$ Ⓒ $\frac{5}{8}$

   Ⓑ $\frac{1}{2}$ Ⓓ $\frac{3}{5}$

4. Wendy has 6 pieces of fruit. Of these, 2 pieces are bananas. What fraction of Wendy's fruit is bananas?
   (Lesson 8.7)

   Ⓐ $\frac{2}{6}$ Ⓒ $\frac{4}{6}$

   Ⓑ $\frac{2}{4}$ Ⓓ $\frac{2}{2}$

5. Toby collects data and makes a bar graph about his classmates' pets. He finds that 9 classmates have dogs, 2 classmates have fish, 6 classmates have cats, and 3 classmates have gerbils. Which pet will have the longest bar on the bar graph? (Lesson 2.5)

   Ⓐ dog   Ⓒ cat

   Ⓑ fish   Ⓓ gerbil

6. The number sentence is an example of which multiplication property? (Lesson 4.4)

   $$6 \times 7 = (6 \times 5) + (6 \times 2)$$

   Ⓐ Associative

   Ⓑ Commutative

   Ⓒ Distributive

   Ⓓ Identity

© Houghton Mifflin Harcourt Publishing Company

Model Equivalent Fractions

Shade the model. Then divide the pieces to find the
equivalent fraction.

1.
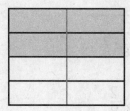

$$\frac{2}{4} = \frac{4}{8}$$

2.
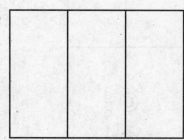

$$\frac{1}{3} = \frac{\square}{6}$$

Use the number line to find the equivalent fraction.

3.
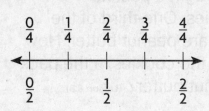

$$\frac{1}{2} = \frac{\square}{4}$$

4.
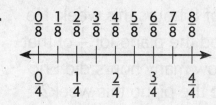

$$\frac{3}{4} = \frac{\square}{8}$$

Problem Solving REAL WORLD

5. Mike says that $\frac{3}{3}$ of his fraction
model is shaded blue. Ryan says that
$\frac{6}{6}$ of the same model is shaded blue.
Are the two fractions equivalent?
If so, what is another equivalent
fraction?

6. Brett shaded $\frac{4}{8}$ of a sheet of
notebook paper. Aisha says he
shaded $\frac{1}{2}$ of the paper. Are the two
fractions equivalent? If so, what is
another equivalent fraction?

© Houghton Mifflin Harcourt Publishing Company

Lesson Check

1. Find the fraction equivalent to $\frac{2}{3}$.

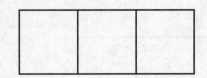

  (A) $\frac{3}{2}$ (C) $\frac{3}{6}$

  (B) $\frac{4}{6}$ (D) $\frac{1}{3}$

2. Find the fraction equivalent to $\frac{1}{4}$.

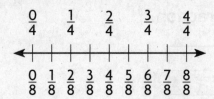

  (A) $\frac{1}{2}$ (C) $\frac{2}{8}$

  (B) $\frac{2}{4}$ (D) $\frac{6}{8}$

Spiral Review

3. Eric practiced piano and guitar for a total of 8 hours this week. He practiced the piano for $\frac{1}{4}$ of that time. How many hours did Eric practice the piano this week? (Lesson 8.8)

  (A) 6 hours   (C) 3 hours

  (B) 4 hours   (D) 2 hours

4. Kylee bought a pack of 12 cookies. One-third of the cookies are peanut butter. How many of the cookies in the pack are peanut butter? (Lesson 8.8)

  (A) 9   (C) 4

  (B) 6   (D) 3

5. There are 56 students going to the game. The coach puts 7 students in each van. Which number sentence can be used to find how many vans are needed to take the students to the game? (Lesson 7.7)

  (A) $56 + 7 = \blacksquare$

  (B) $\blacksquare + 7 = 56$

  (C) $\blacksquare \times 7 = 56$

  (D) $56 - 7 = \blacksquare$

6. Which number sentence can be used to describe the picture? (Lesson 7.1)

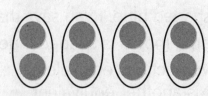

  (A) $2 + 4 = 6$

  (B) $4 - 2 = 2$

  (C) $4 \times 1 = 4$

  (D) $8 \div 2 = 4$

© Houghton Mifflin Harcourt Publishing Company

Equivalent Fractions

Each shape is 1 whole. Shade the model to find the equivalent fraction.

1.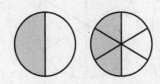

$$\frac{1}{2} = \frac{3}{6}$$

2.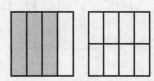

$$\frac{3}{4} = \frac{6}{\boxed{}}$$

Circle equal groups to find the equivalent fraction.

3.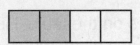

$$\frac{2}{4} = \frac{\boxed{}}{2}$$

4.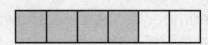

$$\frac{4}{6} = \frac{\boxed{}}{3}$$

Problem Solving REAL WORLD

5. May painted 4 out of 8 equal parts of a poster board blue. Jared painted 2 out of 4 equal parts of a same-size poster board red. Write fractions to show which part of the poster board each person painted.

   _____

6. Are the fractions equivalent? Draw a model to explain.

   _____

© Houghton Mifflin Harcourt Publishing Company

Lesson Check

1. Which fraction is equivalent to $\frac{6}{8}$?

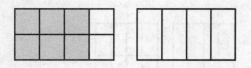

Ⓐ $\frac{1}{4}$ Ⓒ $\frac{3}{4}$

Ⓑ $\frac{1}{3}$ Ⓓ $\frac{4}{6}$

2. Which fraction is equivalent to $\frac{1}{3}$?

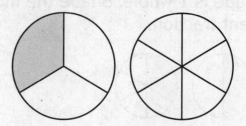

Ⓐ $\frac{1}{6}$ Ⓒ $\frac{2}{6}$

Ⓑ $\frac{2}{8}$ Ⓓ $\frac{2}{3}$

Spiral Review

3. Which number sentence is shown by the array? (Lesson 6.7)

Ⓐ $8 - 2 = 6$

Ⓑ $8 \times 1 = 8$

Ⓒ $2 + 8 = 10$

Ⓓ $16 \div 2 = 8$

4. Cody put 4 plates on the table. He put 1 apple on each plate. Which number sentence can be used to find the total number of apples on the table? (Lesson 3.7)

Ⓐ $4 + 1 = 5$

Ⓑ $4 - 1 = 3$

Ⓒ $4 \times 1 = 4$

Ⓓ $4 \div 2 = 2$

5. Which number sentence is a related fact to $7 \times 3 = 21$?

(Lesson 6.8)

Ⓐ $7 + 3 = 10$

Ⓑ $7 - 3 = 4$

Ⓒ $7 \times 2 = 14$

Ⓓ $21 \div 3 = 7$

6. Find the quotient. (Lesson 7.5)

$$4\overline{)36}$$

Ⓐ 9

Ⓑ 8

Ⓒ 7

Ⓓ 6

© Houghton Mifflin Harcourt Publishing Company

Chapter 9 Extra Practice

Lesson 9.1

Solve. Show your work.

1. Nick finished $\frac{4}{8}$ of his homework before dinner. Ed finished $\frac{7}{8}$ of his homework before dinner. Who finished the greater part of his homework?

_____

2. Rafael walked $\frac{2}{3}$ mile and then rode his scooter $\frac{2}{6}$ mile. Which distance is farther?

_____ mile is farther.

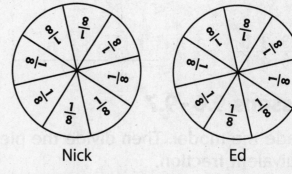

Nick   Ed

$\frac{1}{3}$		$\frac{1}{3}$		$\frac{1}{3}$	
$\frac{1}{6}$	$\frac{1}{6}$	$\frac{1}{6}$	$\frac{1}{6}$	$\frac{1}{6}$	$\frac{1}{6}$

Lessons 9.2 - 9.3

Compare. Write <, >, or =.

1. $\frac{2}{6} \bigcirc \frac{3}{6}$

2. $\frac{6}{8} \bigcirc \frac{1}{8}$

3. $\frac{3}{8} \bigcirc \frac{3}{4}$

4. $\frac{1}{6} \bigcirc \frac{1}{8}$

5. $\frac{2}{3} \bigcirc \frac{2}{6}$

6. $\frac{1}{8} \bigcirc \frac{3}{8}$

Lesson 9.4

Compare. Write <, >, or =. Write the strategy you used.

1. $\frac{2}{8} \bigcirc \frac{2}{3}$

2. $\frac{5}{6} \bigcirc \frac{1}{6}$

3. $\frac{7}{8} \bigcirc \frac{3}{4}$

_____   _____   _____

_____   _____   _____

© Houghton Mifflin Harcourt Publishing Company

Lesson 9.5

Write the fractions in order from greatest to least.

1. $\frac{1}{2}, \frac{1}{4}, \frac{1}{3}$ _____, _____, _____

2. $\frac{4}{6}, \frac{1}{6}, \frac{2}{6}$ _____, _____, _____

Write the fractions in order from least to greatest.

3. $\frac{3}{6}, \frac{3}{4}, \frac{3}{8}$ _____, _____, _____

4. $\frac{6}{8}, \frac{3}{8}, \frac{5}{8}$ _____, _____, _____

Lessons 9.6 - 9.7

Shade the model. Then divide the pieces to find the equivalent fraction.

1.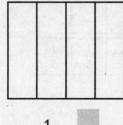

$$\frac{1}{4} = \frac{}{8}$$

2.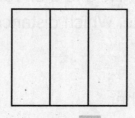

$$\frac{2}{3} = \frac{}{6}$$

Use the number line to find the equivalent fraction.

3.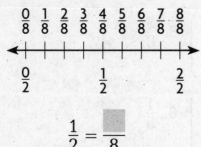

$$\frac{1}{2} = \frac{}{8}$$

4.

$$\frac{2}{2} = \frac{}{6}$$

Each shape is 1 whole. Shade the model to find the equivalent fraction .

5.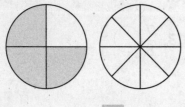

$$\frac{3}{4} = \frac{}{8}$$

6.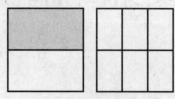

$$\frac{1}{2} = \frac{}{6}$$

© Houghton Mifflin Harcourt Publishing Company

School-Home Letter

Vocabulary

A.M. The times after midnight and before noon

elapsed time The amount of time that passes from the start of an activity to the end of the activity

P.M. The times after noon and before midnight

Dear Family,

During the next few weeks, our math class will be learning about measurement. We will learn to measure time, length, liquid volume, and mass.

You can expect to see homework that provides practice with telling time, finding elapsed time, and solving problems with measurement.

Here is a sample of how your child will be taught to find elapsed time.

🔑 MODEL Find Elapsed Time

This is one way we will be learning to find elapsed time.
Start time: 2:06 P.M. End time: 2:20 P.M.

STEP 1

Find the starting time on a the number line. Count on to the ending time, 2:20.

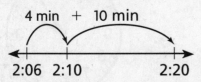

So, the elapsed time is 14 minutes.

STEP 2

Add the minutes.

$$4 + 10 = 14$$

Tips

Another Way to Find Elapsed Time

Another way to find the elapsed time is to use an analog clock.

Activity

Have your child practice telling time and finding elapsed time. Ask questions such as, "Soccer practice starts at 3:30 P.M. It ends at 4:20 P.M. How many minutes does it last?"

© Houghton Mifflin Harcourt Publishing Company

Carta para la casa

Vocabulario

A.M. El tiempo después de la media noche y antes del medio día

tiempo transcurrido El periodo de tiempo que transcurre desde el inicio hasta el final de una actividad

P.M. El tiempo después del medio día y antes de la media noche

Querida familia,

Durante las próximas semanas, en la clase de matemáticas aprenderemos sobre mediciones. Aprenderemos a medir el tiempo, la longitud, el volumen de los líquidos y la masa.

Llevaré a la casa tareas con actividades que incluyen decir la hora, hallar el tiempo transcurrido, y resolver problemas con mediciones.

Este es un ejemplo de la manera como aprenderemos a hallar el tiempo transcurrido.

🔒 MODELO Hallar el tiempo transcurrido

Esta es una manera de hallar el tiempo transcurrido.
Hora de inicio: 2:06 P.M. Final: 2:20 P.M.

PASO 1

Halla en una recta numérica la hora de inicio. Cuenta hacia adelante hasta llegar a la hora final, 2:20.

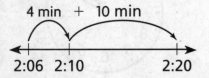

4 min + 10 min

2:06 2:10   2:20

PASO 2

Suma los minutos.

4 + 10 = 14

Por tanto, el tiempo transcurrido es 14 minutos.

Pistas

Otra manera de hallar el tiempo transcurrido

Otra manera de hallar el tiempo transcurrido es usar un reloj analógico.

Actividad

Pida a su hijo o hija que practique cómo decir la hora y hallar el tiempo transcurrido. Usen las actividades familiares o las actividades programadas para practicar el tiempo transcurrido. Por ejemplo, "El entrenamiento de fútbol empieza a las 3:30 P.M. y termina a las 4:20 P.M. ¿Cuántos minutos dura?"

© Houghton Mifflin Harcourt Publishing Company

Name _____

Time to the Minute

Write the time. Write one way you can read the time.

1.

1:16; sixteen minutes after one

2.

_____
_____
_____

3.

4:13

_____
_____
_____

4.

_____
_____
_____

5.

7:24

_____
_____
_____

6.

_____
_____
_____

Write the time another way.

7. 23 minutes after 4

_____

8. 18 minutes before 11

_____

9. 10 minutes before 9

_____

10. 7 minutes after 1

_____

Problem Solving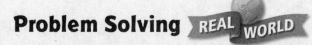

11. What time is it when the hour hand is a little past the 3 and the minute hand is pointing to the 3?

_____

12. Pete began practicing at twenty-five minutes before eight. What is another way to write this time?

_____

© Houghton Mifflin Harcourt Publishing Company

Lesson Check

1. Which is another way to write 13 minutes before 10?

(A) 9:47

(B) 10:13

(C) 10:47

(D) 11:13

2. What time does the clock show?

(A) 2:20   (C) 3:20

(B) 2:40   (D) 4:10

Spiral Review

3. Each bird has 2 wings. How many wings will 5 birds have? (Lesson 3.1)

(A) 7

(B) 8

(C) 9

(D) 10

4. Find the unknown factor. (Lesson 5.2)

$$9 \times \blacksquare = 36$$

(A) 4

(B) 6

(C) 8

(D) 27

5. Mr. Wren has 56 paintbrushes. He places 8 paintbrushes on each of the tables in the art room. How many tables are in the art room?

(Lesson 6.3)

(A) 6

(B) 7

(C) 9

(D) 48

6. Which number completes the equations? (Lesson 6.7)

$$4 \times \blacktriangle = 20 \quad 20 \div 4 = \blacktriangle$$

(A) 4

(B) 5

(C) 6

(D) 16

© Houghton Mifflin Harcourt Publishing Company

Name _____

A.M. and P.M.

Write the time for the activity. Use A.M. or P.M.

1. eat lunch

12:20 P.M.

2. go home after school

_____

3. see the sunrise

_____

4. go for a walk

_____

5. go to school

_____

6. get ready for art class

_____

Write the time. Use A.M. or P.M.

7. 13 minutes after 5:00 in the morning

_____

8. 19 minutes before 9:00 at night

_____

9. quarter before midnight

_____

10. one-half hour after 4:00 in the morning

_____

Problem Solving

11. Jaime is in math class. What time is it? Use A.M. or P.M.

_____

12. Pete began practicing his trumpet at fifteen minutes past three. Write this time using A.M. or P.M.

_____

© Houghton Mifflin Harcourt Publishing Company

Lesson Check

1. Steven is doing his homework. What time is it? Use A.M. or P.M.

- (A) 4:15 P.M.
- (B) 4:25 A.M.
- (C) 4:35 P.M.
- (D) 4:35 A.M.

2. After he finished breakfast, Mr. Edwards left for work at fifteen minutes after seven. What time is this? Use A.M. or P.M.

- (A) 6:15 A.M.
- (B) 7:15 A.M.
- (C) 6:45 P.M.
- (D) 7:30 P.M.

Spiral Review

3. Which division equation is related to the multiplication equation $4 \times 6 = 24$? **(Lesson 6.7)**

- (A) $24 \div 8 = 3$
- (B) $12 \div 3 = 4$
- (C) $6 \times 4 = 24$
- (D) $24 \div 4 = 6$

4. There are 50 toothpicks in each box. Jaime buys 4 boxes for her party platter. How many toothpicks does Jaime buy in all? **(Lesson 5.4)**

- (A) 20
- (B) 54
- (C) 200
- (D) 2,000

5. A pet store sold 145 bags of beef-flavored dog food and 263 bags of cheese-flavored dog food. How many bags of dog food were sold in all? **(Lesson 1.6)**

- (A) 118
- (B) 308
- (C) 408
- (D) 422

6. Victoria and Melody are comparing fraction strips. Which statement is NOT correct? **(Lesson 9.2)**

- (A) $\frac{1}{4} < \frac{4}{4}$
- (B) $\frac{3}{6} > \frac{4}{6}$
- (C) $\frac{2}{8} > \frac{1}{8}$
- (D) $\frac{2}{3} < \frac{3}{3}$

© Houghton Mifflin Harcourt Publishing Company

Measure Time Intervals

Find the elapsed time.

1. Start: 8:10 A.M.   End: 8:45 A.M.

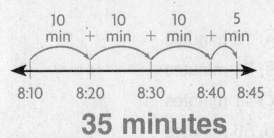

35 minutes

2. Start: 6:45 P.M.   End: 6:54 P.M.

_____

3. Start: 3:00 P.M.   End: 3:37 P.M.

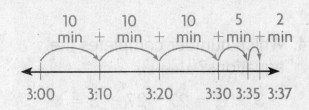

_____

4. Start: 10:05 A.M.   End: 10:21 A.M.

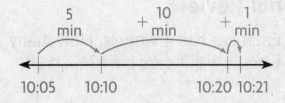

_____

5. Start: 7:30 A.M.   End: 7:53 A.M.

_____

6. Start: 5:20 A.M.   End: 5:47 A.M.

_____

Problem Solving

7. A show at the museum starts at 7:40 P.M. and ends at 7:57 P.M. How long is the show?

_____

8. The first train leaves the station at 6:15 A.M. The second train leaves at 6:55 A.M. How much later does the second train leave the station?

_____

© Houghton Mifflin Harcourt Publishing Company

Lesson Check

1. Marcus began playing basketball at 3:30 P.M. and stopped playing at 3:55 P.M. For how many minutes did he play basketball?

Ⓐ 25 minutes

Ⓑ 30 minutes

Ⓒ 55 minutes

Ⓓ 85 minutes

2. The school play started at 8:15 P.M. and ended at 8:56 P.M. How long was the school play?

Ⓐ 15 minutes

Ⓑ 31 minutes

Ⓒ 41 minutes

Ⓓ 56 minutes

Spiral Review

3. Each car has 4 wheels. How many wheels will 7 cars have? **(Lesson 3.1)**

Ⓐ 11

Ⓑ 24

Ⓒ 27

Ⓓ 28

4. Which number completes the equations? **(Lesson 6.7)**

$$3 \times \blacksquare = 27 \quad 27 \div 3 = \blacksquare$$

Ⓐ 6

Ⓑ 7

Ⓒ 8

Ⓓ 9

5. There are 20 napkins in each package. Kelli bought 8 packages for her party. How many napkins did Kelli buy in all? **(Lesson 5.4)**

Ⓐ 28

Ⓑ 40

Ⓒ 160

Ⓓ 180

6. Mr. Martin drove 290 miles last week. This week he drove 125 miles more than last week. How many miles did Mr. Martin drive this week? **(Lesson 1.7)**

Ⓐ 125 miles

Ⓑ 165 miles

Ⓒ 315 miles

Ⓓ 415 miles

© Houghton Mifflin Harcourt Publishing Company

Use Time Intervals

Find the starting time.

1. Ending time: 4:29 P.M.
   Elapsed time: 55 minutes

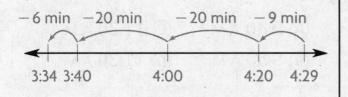

   −6 min   −20 min   −20 min   −9 min

   3:34 3:40   4:00   4:20 4:29

   ### 3:34 P.M.

2. Ending time: 10:08 A.M.
   Elapsed time: 30 minutes

Find the ending time.

3. Starting time: 2:15 A.M.
   Elapsed time: 45 minutes

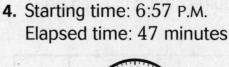

4. Starting time: 6:57 P.M.
   Elapsed time: 47 minutes

Problem Solving

5. Jenny spent 35 minutes doing research on the Internet. She finished at 7:10 P.M. At what time did Jenny start her research?

6. Clark left for school at 7:43 A.M. He got to school 36 minutes later. At what time did Clark get to school?

© Houghton Mifflin Harcourt Publishing Company

Lesson Check

1. Cody and his friends started playing a game at 6:30 P.M. It took them 37 minutes to finish the game. At what time did they finish?

   (A) 5:07 P.M.   (C) 6:53 P.M.

   (B) 5:53 P.M.   (D) 7:07 P.M.

2. Delia worked for 45 minutes on her oil painting. She took a break at 10:35 A.M. At what time did Delia start working on the painting?

   (A) 9:40 A.M.   (C) 11:20 A.M.

   (B) 9:50 A.M.   (D) 11:30 A.M.

Spiral Review

3. Sierra has 30 collector's pins. She wants to put an equal number of pins in each of 5 boxes. How many pins should she put in each box?

   (Lesson 6.4)

   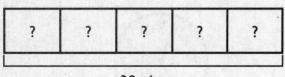

   30 pins

   (A) 4   (C) 6

   (B) 5   (D) 8

4. What time is shown on the clock?

   (Lesson 10.1)

   (A) 1:24   (C) 4:12

   (B) 2:24   (D) 5:12

5. Ricardo has 32 books to put on 4 shelves. He puts the same number of books on each shelf. How many books does Ricardo put on each shelf? (Lesson 7.5)

   (A) 6   (C) 8

   (B) 7   (D) 9

6. Jon started playing a computer game at 5:35 P.M. He finished the game at 5:52 P.M. How long did Jon play the game? (Lesson 10.3)

   (A) 17 minutes   (C) 25 minutes

   (B) 23 minutes   (D) 27 minutes

© Houghton Mifflin Harcourt Publishing Company

Name _____

Problem Solving • Time Intervals

Solve each problem. Show your work.

1. Hannah wants to meet her friends
   downtown. Before leaving home,
   she does chores for 60 minutes and
   eats lunch for 20 minutes. The walk
   downtown takes 15 minutes. Hannah
   starts her chores at 11:45 A.M. At what
   time does she meet her friends?

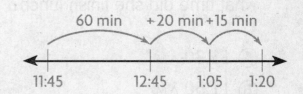

   1:20 P.M.

2. Katie practiced the flute for 45 minutes.
   Then she ate a snack for 15 minutes.
   Next, she watched television for
   30 minutes, until 6:00 P.M. At what time
   did Katie start practicing the flute?

   _____

3. Nick gets out of school at 2:25 P.M. He
   has a 15-minute ride home on the bus.
   Next, he goes on a 30-minute bike
   ride. Then he spends 55 minutes doing
   homework. At what time does Nick
   finish his homework?

   _____

4. The third-grade class is going on a field
   trip by bus to the museum. The bus
   leaves the school at 9:45 A.M. The bus
   ride takes 47 minutes. At what time
   does the bus arrive at the museum?

   _____

© Houghton Mifflin Harcourt Publishing Company

Lesson Check

1. Gloria went to the mall and spent 50 minutes shopping. Then she had lunch for 30 minutes. If Gloria arrived at the mall at 11:00 A.M., at what time did she finish lunch?

   (A) 11:30 A.M.

   (B) 11:50 A.M.

   (C) 12:20 P.M.

   (D) 12:30 P.M.

2. The ball game begins at 2:00 P.M. It takes Ying 30 minutes to get to the ballpark. At what time should Ying leave home to get to the game 30 minutes before it starts?

   (A) 12:30 P.M.

   (B) 1:00 P.M.

   (C) 1:30 P.M.

   (D) 3:00 P.M.

Spiral Review

3. Which lists the fractions in order from least to greatest? (Lesson 9.5)

   (A) $\frac{2}{8}, \frac{2}{4}, \frac{2}{6}$

   (B) $\frac{2}{4}, \frac{2}{8}, \frac{2}{6}$

   (C) $\frac{2}{8}, \frac{2}{6}, \frac{2}{4}$

   (D) $\frac{2}{4}, \frac{2}{6}, \frac{2}{8}$

4. Find the unknown factor. (Lesson 5.2)

   $$6 \times \blacksquare = 36$$

   (A) 4

   (B) 6

   (C) 7

   (D) 8

5. There were 405 books on the library shelf. Some books were checked out. Now there are 215 books left on the shelf. How many books were checked out? (Lesson 1.10)

   (A) 620

   (B) 220

   (C) 210

   (D) 190

6. Savannah has 48 photos. She places 8 photos on each page of her photo album. How many pages in the album does she use? (Lesson 6.3)

   (A) 5

   (B) 6

   (C) 7

   (D) 9

© Houghton Mifflin Harcourt Publishing Company

Measure Length

Measure the length to the nearest half inch.

1.

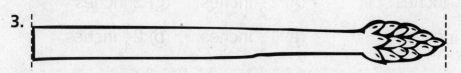

$1\frac{1}{2}$ _____ inches

2.

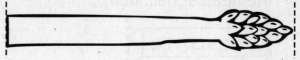

_____ inches

3.

_____ inches

Measure the length to the nearest fourth inch.

4.

_____ inches

5.

_____ inches

6.

_____ inch

7.

_____ inches

Problem Solving REAL WORLD

Use a separate sheet of paper for 8.

8. Draw 8 lines that are between 1 inch and 3 inches long. Measure each line to the nearest fourth inch, and make a line plot.

9. The tail on Alex's dog is $5\frac{1}{4}$ inches long. This length is between which two inch-marks on a ruler?

_____

© Houghton Mifflin Harcourt Publishing Company

Lesson Check

1. What is the length of the eraser to the nearest half inch?

- (A) $\frac{1}{2}$ inch
- (C) $1\frac{1}{2}$ inches
- (B) 1 inch
- (D) 2 inches

2. What is the length of the leaf to the nearest fourth inch?

- (A) $1\frac{1}{2}$ inches
- (C) 2 inches
- (B) $1\frac{3}{4}$ inches
- (D) $2\frac{1}{4}$ inches

Spiral Review

3. Which equation is NOT included in the same set of related facts as $6 \times 8 = 48$? (Lesson 6.8)

- (A) $8 \times 6 = 48$
- (B) $8 \times 8 = 64$
- (C) $48 \div 6 = 8$
- (D) $48 \div 8 = 6$

4. Brooke says there are 49 days until July 4. There are 7 days in a week. In how many weeks will it be July 4? (Lesson 7.7)

- (A) 9 weeks
- (B) 8 weeks
- (C) 7 weeks
- (D) 6 weeks

5. It is 20 minutes before 8:00 in the morning. Which is the correct way to write that time? (Lesson 10.2)

- (A) 7:40 A.M.
- (B) 7:40 P.M.
- (C) 8:20 A.M.
- (D) 8:40 A.M.

6. Marcy played the piano for 45 minutes. She stopped playing at 4:15 P.M. At what time did she start playing the piano? (Lesson 10.4)

- (A) 3:00 P.M.
- (B) 3:30 P.M.
- (C) 4:45 P.M.
- (D) 5:00 P.M.

© Houghton Mifflin Harcourt Publishing Company

Estimate and Measure Liquid Volume

Estimate how much liquid volume there will be when the container is filled. Write *more than 1 liter, about 1 liter,* or *less than 1 liter.*

1. large milk container

more than 1 liter

2. small milk container

3. water bottle

4. spoonful of water

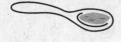

5. bathtub filled halfway

6. filled eyedropper

Problem Solving REAL WORLD

Use the pictures for 7–8. Alan pours water into four glasses that are the same size.

7. Which glass has the most amount of water? _____

8. Which glass has the least amount of water? _____

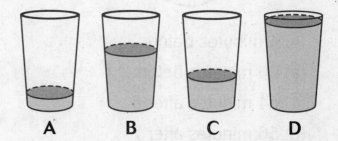

A   B   C   D

© Houghton Mifflin Harcourt Publishing Company

Lesson Check

1. Felicia filled the bathroom sink with water. About how much water does she put in the sink?

   (A) about 1 liter

   (B) more than 1 liter

   (C) a little less than 1 liter

   (D) much less than 1 liter

2. Kyle needed about 1 liter of water to fill a container. Which container did Kyle most likely fill?

   (A) a small glass

   (B) a spoon

   (C) a large pail

   (D) a vase

Spiral Review

3. Cecil had 6 ice cubes. He put 1 ice cube in each glass. In how many glasses did Cecil put ice cubes? (Lesson 6.9)

   (A) 6   (C) 1

   (B) 5   (D) 0

4. Juan has 12 muffins. He puts $\frac{1}{4}$ of the muffins in a bag. How many muffins does Juan put in the bag? (Lesson 8.8)

   (A) 2   (C) 4

   (B) 3   (D) 5

5. Which is one way to read the time shown on the clock? (Lesson 10.1)

   (A) 4 minutes before 7

   (B) 26 minutes before 11

   (C) 54 minutes after 6

   (D) 56 minutes after 7

6. Julianne drew the line segment below. Use your ruler to measure the segment to the nearest fourth inch. (Lesson 10.6)

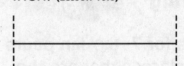

   (A) $\frac{3}{4}$ inch

   (B) $1\frac{1}{4}$ inches

   (C) $1\frac{1}{2}$ inches

   (D) $1\frac{3}{4}$ inches

© Houghton Mifflin Harcourt Publishing Company

Name _____

Estimate and Measure Mass

Choose the unit you would use to measure the mass. Write *gram* or *kilogram*.

1. CD

gram
_____

2. boy

_____

3. bag of sugar

sugar

_____

4. lion

_____

5. paper clip

_____

6. empty plastic bottle

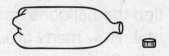

_____

Compare the masses of the objects. Write *is less than*, *is the same as*, or *is more than*.

7.

The mass of the candle _____ the mass of the light bulb.

8.

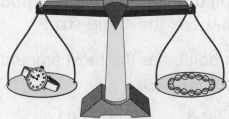

The mass of the watch _____ the mass of the necklace.

Problem Solving REAL WORLD

9. A red ball has a mass that is less than 1 kilogram. A blue ball has a mass of 1 kilogram. Is the mass of the blue ball more than or less than the mass of the red ball?

_____

10. Brock's dog is a collie. To find the mass of his dog, should Brock use *grams* or *kilograms*?

_____

© Houghton Mifflin Harcourt Publishing Company

Lesson Check

1. Which unit of measure would you use to measure the mass of a grape?

ⓐ gram   ⓒ kilogram

ⓑ inch   ⓓ meter

2. Elsie wants to find the mass of her pony. Which unit should she use?

ⓐ gram   ⓒ kilogram

ⓑ liter   ⓓ centimeter

Spiral Review

3. Marsie blew up 24 balloons. She tied the balloons together in groups of 4. How many groups did Marsie make? **(Lesson 6.3)**

ⓐ 5   ⓒ 7

ⓑ 6   ⓓ 8

4. Clark used the order of operations to find the unknown number in $15 - 12 \div 3 = n$. What is the value of the unknown number? **(Lesson 7.11)**

ⓐ 1   ⓒ 9

ⓑ 6   ⓓ 11

Use the pictures for 5–6. Ralph pours juice into four bottles that are the same size.

5. Which bottle has the most amount of juice? **(Lesson 10.7)**

ⓐ Bottle A   ⓒ Bottle C

ⓑ Bottle B   ⓓ Bottle D

6. Which bottle has the least amount of juice? **(Lesson 10.7)**

ⓐ Bottle A   ⓒ Bottle C

ⓑ Bottle B   ⓓ Bottle D

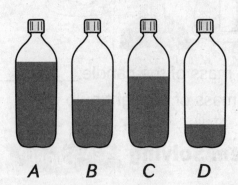

A   B   C   D

© Houghton Mifflin Harcourt Publishing Company

Name _____

Solve Problems About Liquid Volume and Mass

Write an equation and solve the problem.

1. Luis was served 145 grams of meat and 217 grams of vegetables at a meal. What was the total mass of the meat and the vegetables?

   Think: Add to find how much in all.

   145 ⊕ 217 = _____ _____

2. The gas tank of a riding mower holds 5 liters of gas. How many 5-liter gas tanks can you fill from a full 20-liter gas can?

   _____ ◯ _____ = _____ _____

3. To make a lemon-lime drink, Mac mixed 4 liters of lemonade with 2 liters of limeade. How much lemon-lime drink did Mac make?

   _____ ◯ _____ = _____ _____

4. A nickel has a mass of 5 grams. There are 40 nickels in a roll of nickels. What is the mass of a roll of nickels?

   _____ ◯ _____ = _____ _____

5. Four families share a basket of 16 kilograms of apples equally. How many kilograms of apples does each family get?

   _____ ◯ _____ = _____ _____

6. For a party, Julia made 12 liters of fruit punch. There were 3 liters of fruit punch left after the party. How much fruit punch did the people drink at the party?

   _____ ◯ _____ = _____ _____

Problem Solving  REAL WORLD

7. Zoe's fish tank holds 27 liters of water. She uses a 3-liter container to fill the tank. How many times does she have to fill the 3-liter container in order to fill her fish tank?

   _____

8. Adrian's backpack has a mass of 15 kilograms. Theresa's backpack has a mass of 8 kilograms. What is the total mass of both backpacks?

   _____

© Houghton Mifflin Harcourt Publishing Company

Lesson Check

1. Mickey's beagle has a mass of 15 kilograms. His dachshund has a mass of 13 kilograms. What is the combined mass of the two dogs?

(A) 2 kilograms   (C) 23 kilograms

(B) 18 kilograms   (D) 28 kilograms

2. Lois put 8 liters of water in a bucket for her pony. At the end of the day, there were 2 liters of water left. How much water did the pony drink?

(A) 4 liters   (C) 10 liters

(B) 6 liters   (D) 16 liters

Spiral Review

3. Josiah has 3 packs of toy animals. Each pack has the same number of animals. Josiah gives 6 animals to his sister Stephanie. Then Josiah has 9 animals left. How many animals were in each pack? (Lesson 7.10)

(A) 1   (C) 5

(B) 3   (D) 6

4. Tom jogged $\frac{3}{10}$ mile, Betsy jogged $\frac{5}{10}$ mile, and Sue jogged $\frac{2}{10}$ mile. Who jogged a longer distance than $\frac{4}{10}$ mile? (Lesson 9.5)

(A) Betsy

(B) Sue

(C) Tom

(D) None

5. Bob started mowing at 9:55 A.M. It took him 25 minutes to mow the front yard and 45 minutes to mow the back yard. At what time did Bob finish mowing? (Lesson 10.5)

(A) 10:20 A.M.   (C) 11:05 A.M.

(B) 10:55 A.M.   (D) 11:20 A.M.

6. Juliana wants to find the mass of a watermelon. Which unit should she use? (Lesson 10.8)

(A) gram   (C) liter

(B) kilogram   (D) meter

© Houghton Mifflin Harcourt Publishing Company

Chapter 10 Extra Practice

Lessons 10.1 - 10.2

Write the time. Write one way you can read the time.

1.

_____

_____

2.

_____

_____

3.

_____

_____

Write the time. Use A.M. or P.M.

1. 30 minutes past noon

_____

2. 14 minutes before 7:00 in the morning

_____

Lesson 10.3

Find the elapsed time.

1. Start: 10:10 P.M.   End: 10:45 P.M.

⟵————————————⟶

_____

2. Start: 7:05 A.M.   End: 7:33 A.M.

_____

Lessons 10.4 - 10.5

1. Delia spent 45 minutes working on her book report. She finished the report at 6:10 P.M. At what time did Delia start working on her report?

_____

2. Lucas leaves school at 3:05 P.M. The bus ride home takes 25 minutes. Then it takes Lucas 15 minutes to ride his bike to soccer practice. At what time does Lucas get to soccer practice?

_____

© Houghton Mifflin Harcourt Publishing Company

Lesson 10.6

Measure the length to the nearest half inch.

1.

_____ inches

2.

_____ inches

Measure the length to the nearest fourth inch.

3.

_____ inches

Lesson 10.7

Estimate how much liquid volume there will be when the container is filled. Write *more than 1 liter*, *about 1 liter*, or *less than 1 liter*.

1. mug

2. watering can

3. sports bottle

_____

_____

_____

Lesson 10.8

Choose the unit you would use to measure the mass.
Write *gram* or *kilogram*.

1. pen

2. bag of flour

3. brick

_____

_____

_____

Lesson 10.9

Write an equation and solve the problem.

1. Miles ate two hot dogs with buns. Each hot dog has a mass of 45 grams, and each hot dog bun has a mass of 33 grams. How many grams of hot dogs and buns did Miles eat in all?

2. Celia's famous raspberry limeade comes in 3-liter containers. Celia gets an order for 8 containers of raspberry limeade. How many liters of raspberry limeade were ordered?

_____

_____

© Houghton Mifflin Harcourt Publishing Company

School-Home Letter

© Houghton Mifflin Harcourt Publishing Company

Vocabulary

area The measure of unit squares needed to cover a flat surface

perimeter The distance around a shape

unit square A square with a side length of 1 unit that is used to measure area

Dear Family,

During the next few weeks, our math class will be learning about perimeter and area of shapes.

You can expect to see homework that provides practice with measuring and finding perimeter, and finding area by counting squares, using addition, or using multiplication.

Here is a sample of how your child will be taught to find perimeter.

🔑 MODEL Find Perimeter

These are two ways to find perimeter.

Count units.

Find the perimeter of the shape by counting each unit around the shape.

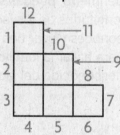

Perimeter is the distance around a shape.

So, the perimeter is 12 units.

Use addition.

Find the perimeter of the rectangle.

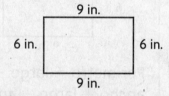

Perimeter = length + width + length + width

Add: $9 + 6 + 9 + 6 = 30$ inches

So, the perimeter is 30 inches.

Tips

Finding Unknown Side Lengths

Sometimes not all lengths of the sides of a shape are given. If you know the perimeter, you can add the lengths of the sides you know and use an equation to find the unknown side length.

Activity

Have your child practice finding the perimeter and area of items around the house. Find and measure the sides of items that have plane shapes, such as an envelope, a place mat, a square potholder, a pennant, or a rug.

Carta
para la casa

Vocabulario

área La medida del número de los cuadrados de una unidad que se necesitan para cubrir una superficie plana

perímetro La distancia alrededor de una figura

cuadrado de una unidad Un cuadrado cuyo lado mide 1 unidad y que se usa para medir un área

unidad cuadrada Una unidad que mide el área como del pies cuadrado, metro cuadrado y así sucesivamente

Querida familia,

Durante las próximas semanas, en la clase de matemáticas aprenderemos acerca del perímetro y el área de las figuras.

Llevaré a la casa tareas que sirven para practicar cómo medir y hallar el perímetro, además de hallar el área contando cuadrados usando la suma o la multiplicación.

Este es un ejemplo de la manera como aprenderemos a hallar el perímetro.

🔒 MODELO Hallar el perímetro

Estas son dos maneras de hallar el perímetro.

Contar unidades.

Halla el perímetro de la figura contando cada unidad alrededor de la figura.

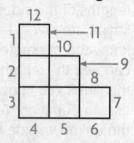

El perímetro es la distancia alrededor de una figura.

Por tanto, el perímetro es 12 unidades.

Usar la suma.

Halla el perímetro del rectángulo.

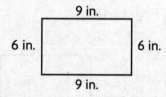

Perímetro = largo + ancho + largo + ancho

Sumo: 9 + 6 + 9 + 6 = 30 pulgadas

Por tanto, el perímetro es 30 pulgadas.

Pistas

Hallar longitudes desconocidas de los lados

A veces no se dan las longitudes de los lados de una figura. Si conoces el perímetro, puedes sumar las longitudes de los lados que conoces y usar una ecuación para hallar la longitud desconocida del lado.

Actividad

Pida a su hijo que practique hallando el perímetro y el área de algunos objetos de la casa. Hallen y midan los lados de objetos que tengan formas planas, como un sobre, un individual para la mesa, un agarrador de ollas cuadrado, un banderín o un tapete.

© Houghton Mifflin Harcourt Publishing Company

Model Perimeter

Find the perimeter of the shape. Each unit is 1 centimeter.

1.
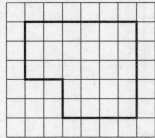

_____ **22** _____ centimeters

2.
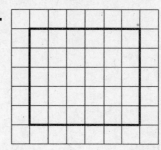

_____ centimeters

3.
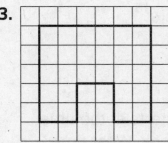

_____ centimeters

4.
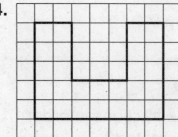

_____ centimeters

Problem Solving REAL WORLD

Use the drawing for 5–6. Each unit is 1 centimeter.

5. What is the perimeter of Patrick's shape?

_____

6. How much greater is the perimeter of Jillian's shape than the perimeter of Patrick's shape?

_____

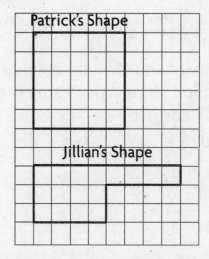

Patrick's Shape

Jillian's Shape

© Houghton Mifflin Harcourt Publishing Company

Lesson Check

1. Find the perimeter of the shape. Each unit is 1 centimeter.

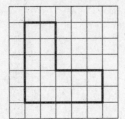

- (A) 14 centimeters
- (B) 16 centimeters
- (C) 18 centimeters
- (D) 20 centimeters

2. Find the perimeter of the shape. Each unit is 1 centimeter.

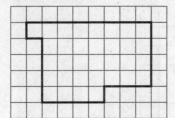

- (A) 19 centimeters
- (B) 26 centimeters
- (C) 33 centimeters
- (D) 55 centimeters

Spiral Review

3. Which lists the fractions in order from least to greatest? (Lesson 9.5)

$$\frac{2}{4}, \frac{2}{3}, \frac{2}{6}$$

- (A) $\frac{2}{3}, \frac{2}{4}, \frac{2}{6}$
- (B) $\frac{2}{6}, \frac{2}{4}, \frac{2}{3}$
- (C) $\frac{2}{4}, \frac{2}{3}, \frac{2}{6}$
- (D) $\frac{2}{3}, \frac{2}{6}, \frac{2}{4}$

4. Kasey's school starts at the time shown on the clock. What time does Kasey's school start? (Lesson 10.1)

- (A) 6:40
- (C) 8:30
- (B) 8:06
- (D) 9:30

5. Michael and Dex are comparing fraction strips. Which statement is NOT correct? (Lesson 9.2)

- (A) $\frac{1}{2} < \frac{2}{2}$
- (C) $\frac{4}{8} < \frac{3}{8}$
- (B) $\frac{2}{3} > \frac{1}{3}$
- (D) $\frac{4}{6} > \frac{2}{6}$

6. Aiden wants to find the mass of a bowling ball. Which unit should he use? (Lesson 10.8)

- (A) liter
- (C) gram
- (B) inch
- (D) kilogram

© Houghton Mifflin Harcourt Publishing Company

Find Perimeter

Use a ruler to find the perimeter.

1.

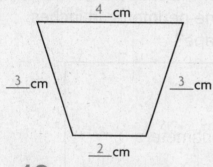
___4___ cm
___3___ cm   ___3___ cm
___2___ cm

12 centimeters

2.

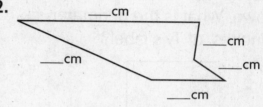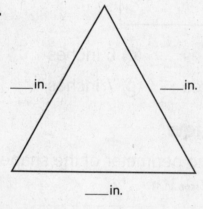
_____ cm
_____ cm
_____ cm
_____ cm
_____ cm

_____ centimeters

3.

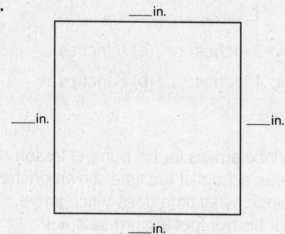
_____ in.
_____ in.   _____ in.
_____ in.

_____ inches

4.

_____ in.   _____ in.
_____ in.

_____ inches

Problem Solving REAL WORLD

Draw a picture to solve 5–6.

5. Evan has a square sticker that measures 5 inches on each side. What is the perimeter of the sticker?

_____

6. Sophie draws a shape that has 6 sides. Each side is 3 centime____ What is the perimeter of the ____

_____

© Houghton Mifflin Harcourt Publishing Company

Lesson Check

Use an inch ruler for 1–2.

1. Ty cut a label the size of the shape shown. What is the perimeter, in inches, of Ty's label?

- (A) 4 inches
- (C) 6 inches
- (B) 5 inches
- (D) 7 inches

2. Julie drew the shape shown below. What is the perimeter, in inches, of the shape?

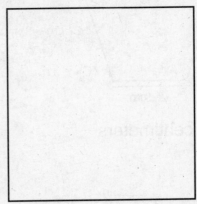

- (A) 2 inches
- (C) 6 inches
- (B) 4 inches
- (D) 8 inches

Spiral Review

3. What is the perimeter of the shape below? **(Lesson 11.1)**

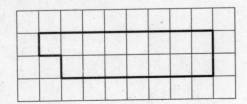

- (A) 8 units
- (C) 20 units
- (B) 10 units
- (D) 22 units

4. Vince arrives for his trumpet lesson after school at the time shown on the clock. What time does Vince arrive for his trumpet lesson? **(Lesson 10.2)**

- (A) 3:26 A.M.
- (B) 4:26 A.M.
- (C) 3:26 P.M.
- (D) 4:26 P.M.

5. Matthew's small fish tank holds 12 liters. His large fish tank holds 25 liters. How many more liters does his large fish tank hold?

(Lesson 10.9)

- (A) 12 liters
- (C) 25 liters
- (B) 13 liters
- (D) 37 liters

6. Cecila and Sasha are comparing fraction strips. Which statement is correct? **(Lesson 9.3)**

- (A) $\frac{1}{2} < \frac{1}{3}$
- (C) $\frac{1}{4} > \frac{1}{2}$
- (B) $\frac{1}{8} > \frac{1}{6}$
- (D) $\frac{1}{6} < \frac{1}{4}$

© Houghton Mifflin Harcourt Publishing Company

Name _____

Find Unknown Side Lengths

Find the unknown side lengths.

1. Perimeter = 33 centimeters

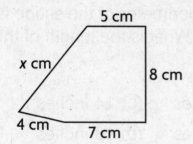

$5 + 8 + 7 + 4 + x = 33$
$24 + x = 33$
$x = 9$

_____**9**_____ centimeters

2. Perimeter = 14 feet

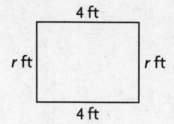

_____ feet

3. Perimeter = 37 meters

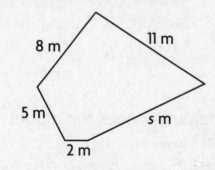

_____ meters

4. Perimeter = 92 inches

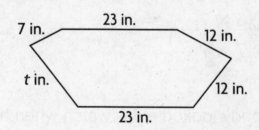

_____ inches

Problem Solving

5. Steven has a rectangular rug with a perimeter of 16 feet. The width of the rug is 5 feet. What is the length of the rug?

_____

6. Kerstin has a square tile. The perimeter of the tile is 32 inches. What is the length of each side of the tile?

_____

© Houghton Mifflin Harcourt Publishing Company

Lesson Check

1. Jesse is putting a ribbon around a square frame. He uses 24 inches of ribbon. How long is each side of the frame?

   (A) 4 inches

   (B) 5 inches

   (C) 6 inches

   (D) 8 inches

2. Davia draws a shape with 5 sides. Two sides are each 5 inches long. Two other sides are each 4 inches long. The perimeter of the shape is 27 inches. What is the length of the fifth side?

   (A) 9 inches   (C) 14 inches

   (B) 13 inches   (D) 18 inches

Spiral Review

3. Which of the following represents $7 + 7 + 7 + 7$? (Lesson 3.2)

   (A) 4×4

   (B) 4×7

   (C) 6×7

   (D) 7×7

4. Bob bought 3 packs of model cars. He gave 4 cars to Ann. Bob has 11 cars left. How many model cars were in each pack? (Lesson 7.10)

   (A) 18   (C) 7

   (B) 11   (D) 5

5. Randy looked at his watch when he started and finished reading. How long did Randy read? (Lesson 10.3)

Start

End

   (A) 55 minutes   (C) 35 minutes

   (B) 45 minutes   (D) 15 minutes

6. Which statement does the model represent? (Lesson 8.6)

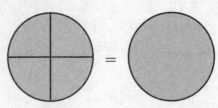

   (A) $\frac{4}{4} = 1$ (C) $\frac{2}{4} = 1$

   (B) $\frac{3}{4} = 1$ (D) $\frac{1}{4} = 1$

© Houghton Mifflin Harcourt Publishing Company

Name _____

Understand Area

Count to find the area for the shape.

1.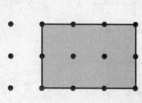

Area = __6__ square units

2.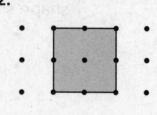

Area = _____ square units

3.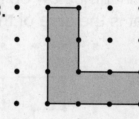

Area = _____ square units

4.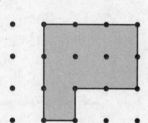

Area = _____ square units

5.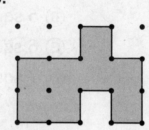

Area = _____ square units

6.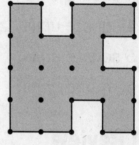

Area = _____ square units

Write *area* or *perimeter* for each situation.

7. carpeting a floor

8. fencing a garden

_____   _____

Problem Solving REAL WORLD

Use the diagram for 9–10.

9. Roberto is building a platform for his model railroad. What is the area of the platform?

_____

10. Roberto will put a border around the edges of the platform. How much border will he need?

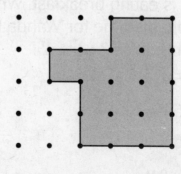

_____

© Houghton Mifflin Harcourt Publishing Company

Lesson Check

1. Josh used rubber bands to make the shape below on his geoboard. What is the area of the shape?

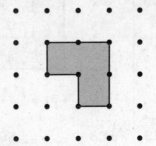

Ⓐ 3 square units

Ⓑ 4 square units

Ⓒ 5 square units

Ⓓ 6 square units

2. Wilma drew the shape below on dot paper. What is the area of the shape she drew?

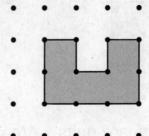

Ⓐ 4 square units

Ⓑ 5 square units

Ⓒ 6 square units

Ⓓ 7 square units

Spiral Review

3. Leonardo knows it is 42 days until summer break. How many weeks is it until Leonardo's summer break? (Hint: There are 7 days in a week.) **(Lesson 7.7)**

Ⓐ 5 weeks   Ⓒ 7 weeks

Ⓑ 6 weeks   Ⓓ 8 weeks

4. Nan cut a submarine sandwich into 4 equal parts and ate one part. What fraction represents the part of the sandwich Nan ate? **(Lesson 8.3)**

Ⓐ $\frac{1}{4}$ Ⓒ $\frac{4}{4}$

Ⓑ $\frac{1}{3}$ Ⓓ $\frac{4}{1}$

5. Wanda is eating breakfast. Which is a reasonable time for Wanda to be eating breakfast? **(Lesson 10.2)**

Ⓐ 7:45 A.M.

Ⓑ 7:45 P.M.

Ⓒ 2:15 A.M.

Ⓓ 2:15 P.M.

6. Dick has 2 bags of dog food. Each bag contains 5 kilograms of food. How many kilograms of food does Dick have in all? **(Lesson 10.8)**

Ⓐ 3 kilograms   Ⓒ 7 kilograms

Ⓑ 5 kilograms   Ⓓ 10 kilograms

© Houghton Mifflin Harcourt Publishing Company

Measure Area

Count to find the area of the shape.
Each unit square is 1 square centimeter.

1.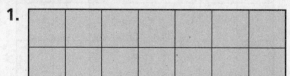

Area = __14__ square centimeters

2.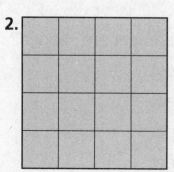

Area = _____ square centimeters

3.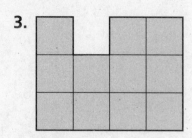

Area = _____ square centimeters

4.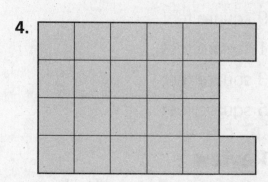

Area = _____ square centimeters

Problem Solving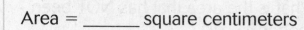

Alan is painting his deck gray. Use
the diagram at the right for 5–6. Each
unit square is 1 square meter.

Alan's Deck

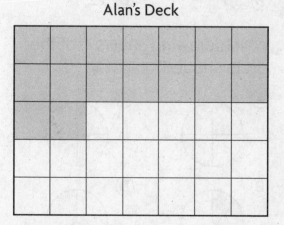

5. What is the area of the deck that
   Alan has already painted gray?

   _____

6. What is the area of the deck that
   Alan has left to paint?

   _____

© Houghton Mifflin Harcourt Publishing Company

Lesson Check

Each unit square in the diagram is
1 square foot.

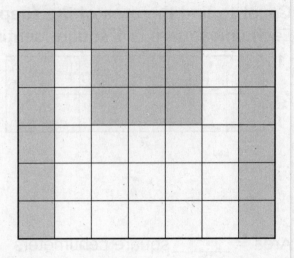

1. How many square feet are shaded?

   Ⓐ 19 square feet

   Ⓑ 21 square feet

   Ⓒ 23 square feet

   Ⓓ 25 square feet

2. What is the area that has NOT been
   shaded?

   Ⓐ 19 square feet

   Ⓑ 21 square feet

   Ⓒ 23 square feet

   Ⓓ 25 square feet

Spiral Review

3. Sonya buys 6 packages of rolls.
   There are 6 rolls in each package.
   How many rolls does Sonya buy?
   (Lesson 4.3)

   Ⓐ 42   Ⓒ 24

   Ⓑ 36   Ⓓ 12

4. Charlie mixed 6 liters of juice with
   2 liters of soda to make fruit punch.
   How many liters of fruit punch did
   Charlie make? (Lesson 10.9)

   Ⓐ 3 liters   Ⓒ 8 liters

   Ⓑ 4 liters   Ⓓ 12 liters

5. Which drawing shows $\frac{2}{3}$ of the
   circle shaded? (Lesson 8.4)

   Ⓐ   Ⓒ

   Ⓑ   Ⓓ

6. Use the models to name a fraction
   that is equivalent to $\frac{1}{2}$. (Lesson 9.7)

   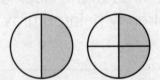

   Ⓐ $\frac{2}{1}$ Ⓒ $\frac{2}{4}$

   Ⓑ $\frac{2}{2}$ Ⓓ $\frac{4}{4}$

© Houghton Mifflin Harcourt Publishing Company

Name _____

Use Area Models

Find the area of each shape. Each unit square is 1 square foot.

1.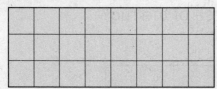

2.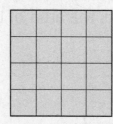

There are 3 rows of 8 unit squares.
$3 \times 8 = 24$

24 square feet
_____   _____

**Find the area of each shape.
Each unit square is 1 square meter.**

3.

4.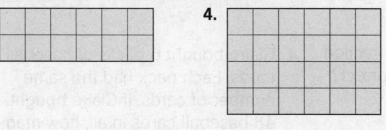

5.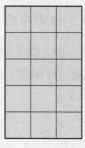

_____   _____   _____

Problem Solving REAL WORLD

6. Landon made a rug for the hallway. Each unit square is 1 square foot. What is the area of the rug?

_____

7. Eva makes a border at the top of a picture frame. Each unit square is 1 square inch. What is the area of the border?

_____

© Houghton Mifflin Harcourt Publishing Company

Lesson Check

1. The entrance to an office has a tiled floor. Each square tile is 1 square meter. What is the area of the floor?

   (A) 8 square meters

   (B) 9 square meters

   (C) 10 square meters

   (D) 12 square meters

2. Ms. Burns buys a new rug. Each unit square is 1 square foot. What is the area of the rug?

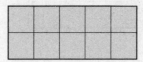

   (A) 5 square feet

   (B) 7 square feet

   (C) 10 square feet

   (D) 12 square feet

Spiral Review

3. Ann and Bill are comparing fraction strips. Which statement is correct?

   (Lesson 9.2)

   (A) $\frac{3}{8} > \frac{5}{8}$

   (B) $\frac{3}{4} < \frac{1}{4}$

   (C) $\frac{3}{6} > \frac{4}{6}$

   (D) $\frac{1}{3} < \frac{2}{3}$

4. Claire bought 6 packs of baseball cards. Each pack had the same number of cards. If Claire bought 48 baseball cards in all, how many cards were in each pack? (Lesson 7.8)

   (A) 54   (C) 8

   (B) 42   (D) 6

5. Austin left for school at 7:35 A.M.. He arrived at school 15 minutes later. What time did Austin arrive at school? (Lesson 10.4)

   (A) 7:40 A.M.   (C) 7:55 A.M.

   (B) 7:50 A.M.   (D) 8:00 A.M.

6. Wyatt's room is a rectangle with a perimeter of 40 feet. The width of the room is 8 feet. What is the length of the room? (Lesson 11.3)

   (A) 5 feet   (C) 16 feet

   (B) 12 feet   (D) 32 feet

© Houghton Mifflin Harcourt Publishing Company

Name _____

Problem Solving • Area of Rectangles

Use the information for 1–3.

An artist makes rectangular murals in different sizes. Below are the available sizes. Each unit square is 1 square meter.

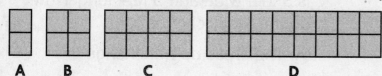

A   B   C   D

1. Complete the table to find the area of each mural.

Mural	Length (in meters)	Width (in meters)	Area (in square meters)
A	2	1	2
B	2	2	4
C	2		
D	2		

2. Find and describe a pattern of how the length changes and how the width changes for murals A through D.

_____

_____

3. How do the areas of the murals change when the width changes?

_____

4. Dan built a deck that is 5 feet long and 5 feet wide. He built another deck that is 5 feet long and 7 feet wide. He built a third deck that is 5 feet long and 9 feet wide. How do the areas change?

_____

© Houghton Mifflin Harcourt Publishing Company

Lesson Check

1. Lauren drew the designs below. Each unit square is 1 square centimeter. If the pattern continues, what will be the area of the fourth shape?

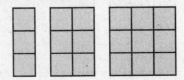

Ⓐ 10 square centimeters

Ⓑ 12 square centimeters

Ⓒ 14 square centimeters

Ⓓ 16 square centimeters

2. Henry built one garden that is 3 feet wide and 3 feet long. He also built a garden that is 3 feet wide and 6 feet long, and a garden that is 3 feet wide and 9 feet long. How do the areas change?

Ⓐ The areas do not change.

Ⓑ The areas double.

Ⓒ The areas increase by 3 square feet.

Ⓓ The areas increase by 9 square feet.

Spiral Review

3. Joe, Jim, and Jack share 27 football cards equally. How many cards does each boy get? (Lesson 7.4)

Ⓐ 7

Ⓑ 8

Ⓒ 9

Ⓓ 10

4. Nita uses $\frac{1}{3}$ of a carton of 12 eggs. How many eggs does she use?

(Lesson 8.7)

Ⓐ 3   Ⓒ 6

Ⓑ 4   Ⓓ 9

5. Brenda made 8 necklaces. Each necklace has 10 large beads. How many large beads did Brenda use to make the necklaces? (Lesson 5.4)

Ⓐ 80

Ⓑ 85

Ⓒ 90

Ⓓ 100

6. Neal is tiling his kitchen floor. Each square tile is 1 square foot. Neal uses 6 rows of tiles with 9 tiles in each row. What is the area of the floor? (Lesson 11.6)

Ⓐ 15 square feet

Ⓑ 52 square feet

Ⓒ 54 square feet

Ⓓ 57 square feet

© Houghton Mifflin Harcourt Publishing Company

Name _____

Area of Combined Rectangles

Use the Distributive Property to find the area.
Show your multiplication and addition equations.

1.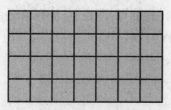

$4 \times 2 = 8$, $4 \times 5 = 20$

$8 + 20 = 28$

___28___ square units

2.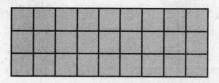

_____

_____

_____ square units

Draw a line to break apart the shape into
rectangles. Find the area of the shape.

3.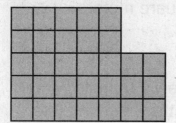

Rectangle 1: _____ × _____ = _____

Rectangle 2: _____ × _____ = _____

_____ + _____ = _____ square units

4.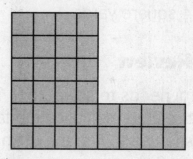

Rectangle 1: _____ × _____ = _____

Rectangle 2: _____ × _____ = _____

_____ + _____ = _____ square units

Problem Solving REAL WORLD

A diagram of Frank's room is at right.
Each unit square is 1 square foot.

5. Draw a line to divide the shape of
Frank's room into rectangles.

6. What is the total area of Frank's room?

_____ square feet

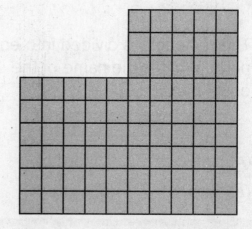

© Houghton Mifflin Harcourt Publishing Company

Lesson Check

1. The diagram shows Ben's backyard. Each unit square is 1 square yard. What is the area of Ben's backyard?

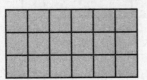

Ⓐ 12 square yards

Ⓑ 16 square yards

Ⓒ 18 square yards

Ⓓ 24 square yards

2. The diagram shows a room in an art gallery. Each unit square is 1 square meter. What is the area of the room?

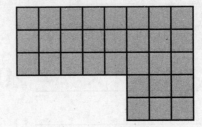

Ⓐ 24 square meters

Ⓑ 30 square meters

Ⓒ 36 square meters

Ⓓ 40 square meters

Spiral Review

3. Naomi needs to solve $28 \div 7 = \blacksquare$. What related multiplication fact can she use to find the unknown number? (Lesson 6.7)

Ⓐ $3 \times 7 = 21$

Ⓑ $4 \times 7 = 28$

Ⓒ $5 \times 7 = 35$

Ⓓ $6 \times 7 = 42$

4. Karen drew a triangle with side lengths 3 centimeters, 4 centimeters, and 5 centimeters. What is the perimeter of the triangle? (Lesson 11.2)

Ⓐ 7 centimeters

Ⓑ 9 centimeters

Ⓒ 11 centimeters

Ⓓ 12 centimeters

5. The rectangle is divided into equal parts. What is the name of the equal parts? (Lesson 8.1)

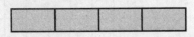

Ⓐ half   Ⓒ fourth

Ⓑ third   Ⓓ sixth

6. Use an inch ruler. To the nearest half inch, how long is this line segment? (Lesson 10.6)

Ⓐ 1 inch   Ⓒ 2 inches

Ⓑ $1\frac{1}{2}$ inches   Ⓓ $2\frac{1}{2}$ inches

© Houghton Mifflin Harcourt Publishing Company

Same Perimeter, Different Areas

Find the perimeter and the area.
Tell which rectangle has a greater area.

1.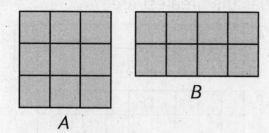

2.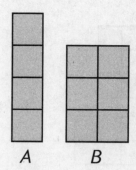

A: Perimeter = ___**12 units**___ ;

Area = ___**9 square units**___

B: Perimeter = _____ ;

Area = _____

Rectangle _____ has a greater area.

A: Perimeter = _____ ;

Area = _____

B: Perimeter = _____ ;

Area = _____

Rectangle _____ has a greater area.

Problem Solving REAL WORLD

3. Tara's and Jody's bedrooms are shaped like rectangles. Tara's bedroom is 9 feet long and 8 feet wide. Jody's bedroom is 7 feet long and 10 feet wide. Whose bedroom has the greater area? **Explain**.

_____

_____

_____

4. Mr. Sanchez has 16 feet of fencing to put around a rectangular garden. He wants the garden to have the greatest possible area. How long should the sides of the garden be?

_____

_____

© Houghton Mifflin Harcourt Publishing Company

Lesson Check

1. Which shape has a perimeter of 12 units and an area of 8 square units?

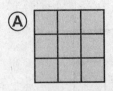
Ⓐ

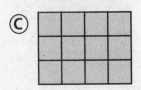
Ⓑ

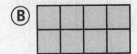
Ⓒ

Ⓓ

2. All four rectangles below have the same perimeter. Which rectangle has the greatest area?

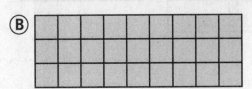
Ⓐ

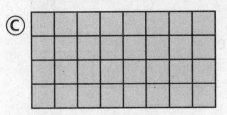
Ⓑ

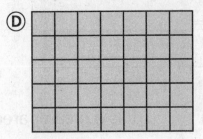
Ⓒ

Ⓓ

Spiral Review

3. Kerrie covers a table with 8 rows of square tiles. There are 7 tiles in each row. What is the area that Kerrie covers in square units?

(Lesson 11.6)

Ⓐ 15 square units

Ⓑ 35 square units

Ⓒ 42 square units

Ⓓ 56 square units

4. Von has a rectangular workroom with a perimeter of 26 feet. The length of the workroom is 6 feet. What is the width of Von's workroom? (Lesson 11.3)

Ⓐ 7 feet

Ⓑ 13 feet

Ⓒ 20 feet

Ⓓ 26 feet

© Houghton Mifflin Harcourt Publishing Company

Same Area, Different Perimeters

Find the perimeter and the area. Tell which
rectangle has a greater perimeter.

1.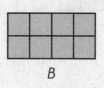

A: Area = __**8 square units**__ ;

Perimeter = __**18 units**__

B: Area = _____ ;

Perimeter = _____

Rectangle _____ has a greater perimeter.

2.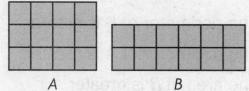

A: Area = _____ ;

Perimeter = _____

B: Area = _____ ;

Perimeter = _____

Rectangle _____ has a greater perimeter.

3.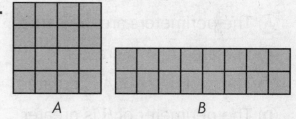

A: Area = _____ ;

Perimeter = _____

B: Area = _____ ;

Perimeter = _____

Rectangle _____ has a greater perimeter.

Problem Solving

Use the tile designs for 4–5.

4. Compare the areas of Design A
   and Design B.

   _____

   _____

5. Compare the perimeters. Which
   design has the greater perimeter?

   _____

Beth's Tile Designs

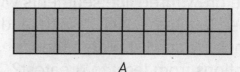

© Houghton Mifflin Harcourt Publishing Company

Lesson Check

1. Jake drew two rectangles. Which statement is true?

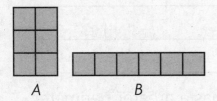

- (A) The perimeters are the same.
- (B) The area of *A* is greater.
- (C) The perimeter of *A* is greater.
- (D) The perimeter of *B* is greater.

2. Alyssa drew two rectangles. Which statement is true?

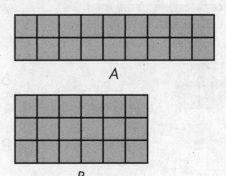

- (A) The perimeter of *B* is greater.
- (B) The perimeter of *A* is greater.
- (C) The area of *B* is greater.
- (D) The perimeters are the same.

Spiral Review

3. Marsha was asked to find the value of $8 - 3 \times 2$. She wrote a wrong answer. Which is the correct answer? (Lesson 7.11)

- (A) 22
- (C) 4
- (B) 10
- (D) 2

4. What fraction names the point on the number line? (Lesson 8.5)

- (A) $\frac{1}{4}$
- (C) $\frac{3}{4}$
- (B) $\frac{2}{3}$
- (D) $\frac{3}{1}$

5. Kyle drew three line segments with these lengths: $\frac{2}{4}$ inch, $\frac{2}{3}$ inch, and $\frac{2}{6}$ inch. Which list orders the fractions from least to greatest?

(Lesson 9.5)

- (A) $\frac{2}{6}, \frac{2}{4}, \frac{2}{3}$
- (C) $\frac{2}{4}, \frac{2}{3}, \frac{2}{6}$
- (B) $\frac{2}{3}, \frac{2}{4}, \frac{2}{6}$
- (D) $\frac{2}{6}, \frac{2}{3}, \frac{2}{4}$

6. On Monday, $\frac{3}{8}$ inch of snow fell. On Tuesday, $\frac{5}{8}$ inch of snow fell. Which statement correctly compares the snow amounts? (Lesson 9.2)

- (A) $\frac{3}{8} = \frac{5}{8}$
- (C) $\frac{5}{8} < \frac{3}{8}$
- (B) $\frac{3}{8} < \frac{5}{8}$
- (D) $\frac{3}{8} > \frac{5}{8}$

© Houghton Mifflin Harcourt Publishing Company

Name _____

Chapter 11 Extra Practice

Lessons 11.1, 11.3

1. Find the perimeter of the shape. Each unit is 1 centimeter.

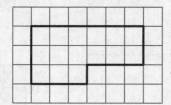

2. The square has a perimeter of 28 inches. What is the length of each side of the square?

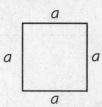

_____   _____

Lesson 11.2

Use a centimeter ruler to find the perimeter.

1.

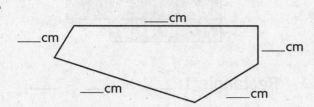

2.

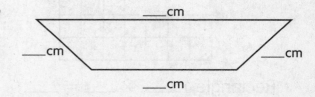

_____   _____

Lessons 11.4 - 11.6

Find the area of the shape.
Each unit square is 1 square inch.

1.

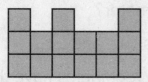

2.

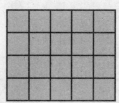

Area = _____ square inches   _____

© Houghton Mifflin Harcourt Publishing Company

Lesson 11.7

Use the rectangles at the right for 1–2.

1. How do the length and width change
   from Rectangle *A* to Rectangle *B*?

   _____

   _____

2 ft

2 ft

A

2. How do the areas change from Rectangle
   A to Rectangle *B* to Rectangle *C*?

   _____

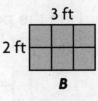

3 ft

2 ft

B

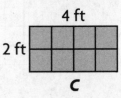

4 ft

2 ft

C

Lesson 11.8

Draw a line to break apart the shape into rectangles.
Find the area of the shape.

1.

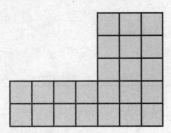

Rectangle 1: ___ × ___ = ___;

Rectangle 2: ___ × ___ = ___;

___ + ___ = ___ square units

2.

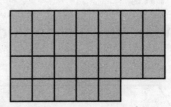

Rectangle 1: ___ × ___ = ___;

Rectangle 2: ___ × ___ = ___;

___ + ___ = ___ square units

Lessons 11.9 - 11.10

Find the perimeter and area of each rectangle.
Use your results to answer questions 1–2.

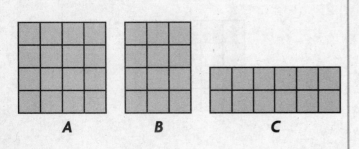

A   **B**   **C**

1. Which two rectangles have the
   same perimeter?

   Rectangles ___ and ___

2. Which two rectangles have the
   same area?

   Rectangles ___ and ___

© Houghton Mifflin Harcourt Publishing Company

School-Home Letter

© Houghton Mifflin Harcourt Publishing Company

Vocabulary

angle A shape formed by two rays that share an endpoint

closed shape A shape that begins and ends at the same point

polygon A closed plane shape made up of straight line segments

quadrilateral A polygon with four sides and four angles

Dear Family,

During the next few weeks, our math class will be learning about plane shapes. We will learn to identify polygons and describe them by their sides and angles.

You can expect to see homework that provides practice with shapes.

Here is a sample of how your child will be taught to classify quadrilaterals.

🔑 MODEL Classify Quadrilaterals

Use sides and angles to name this quadrilateral.

STEP 1 There are 2 right angles.

STEP 2 There is exactly 1 pair of opposite sides that are parallel.

So, the quadrilateral is a trapezoid.

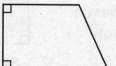

Tips

Checking Angles

The corner of a sheet of paper or an index card can be used to check whether an angle in a polygon is *right, less than a right angle*, or *greater than a right angle*.

Activity

Point out everyday objects that resemble plane shapes, such as books, photos, windows, and traffic signs. Have your child identify the shape and describe it by its sides and angles.

Carta
para la casa

Vocabulario

ángulo Una figura compuesta por dos rayos que comparten un extremo

figura cerrada Una figura que comienza y termina en el mismo punto

polígono Una figura plana cerrada compuesta por segmentos rectos

cuadrilátero Un polígono con cuatro lados y cuatro ángulos

Querida familia,

Durante las próximas semanas, en la clase de matemáticas aprenderemos sobre figuras planas. Aprenderemos a identificar polígonos y a describirlos según sus lados y ángulos.

Llevaré a casa tareas para practicar con figuras.

Este es un ejemplo de cómo clasificaremos cuadriláteros.

🔑 MODELO Clasificar cuadriláteros

Usa los lados y los ángulos para nombrar este cuadrilátero.

PASO 1 Hay dos ángulos rectos.

PASO 2 Hay exactamente 1 par de lados opuestos que son paralelos.

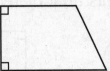

Por tanto, el cuadrilátero es un trapecio.

Pistas

Comprobar ángulos

Puedes usar la esquina de una hoja o de una tarjeta para comprobar si un ángulo de un polígono es *recto, menor que un ángulo recto* o *mayor que un ángulo recto.*

Actividad

Señalen objetos cotidianos que parezcan figuras planas, como libros, fotografías, ventanas y señales de tráfico. Pida a su hijo o hija que identifique la figura y que la describa según sus lados y ángulos.

© Houghton Mifflin Harcourt Publishing Company

Name _____

Describe Plane Shapes

Write how many line segments the shape has.

1.

__4__ line segments

2.

_____ line segments

3.

_____ line segments

4.

_____ line segments

Write whether the shape is *open* or *closed*.

5.

_____

6.

_____

Problem Solving REAL WORLD

7. Carl wants to show a closed shape in his drawing. Show and explain how to make the drawing a closed shape.

_____

_____

8. The shape of a fish pond at a park is shown below. Is the shape open or closed?

_____

_____

© Houghton Mifflin Harcourt Publishing Company

Lesson Check

1. How many line segments does this shape have?

Ⓐ 2   Ⓒ 4

Ⓑ 3   Ⓓ 5

2. Which of these is part of a line, has one endpoint, and continues in one direction?

Ⓐ ray

Ⓑ line

Ⓒ line segment

Ⓓ point

Spiral Review

3. What multiplication sentence does the array show? (Lesson 3.5)

Ⓐ $3 \times 8 = 24$ Ⓒ $8 \times 5 = 40$

Ⓑ $4 \times 8 = 32$ Ⓓ $4 \times 9 = 36$

4. What is the unknown factor and quotient? (Lesson 6.8)

$9 \times \boxed{} = 27$

$27 \div 9 = \boxed{}$

Ⓐ 3

Ⓑ 4

Ⓒ 5

Ⓓ 6

5. Which fraction is equivalent to $\frac{4}{8}$? (Lesson 9.6)

Ⓐ $\frac{3}{4}$ Ⓒ $\frac{1}{4}$

Ⓑ $\frac{1}{2}$ Ⓓ $\frac{1}{8}$

6. Mr. MacTavish has 30 students from his class going on a field trip to the zoo. He is placing 6 students in each group. How many groups of students from Mr. MacTavish's class will be going to the zoo? (Lesson 7.6)

Ⓐ 5   Ⓒ 7

Ⓑ 6   Ⓓ 36

© Houghton Mifflin Harcourt Publishing Company

Name _____

Describe Angles in Plane Shapes

Use the corner of a sheet of paper to tell whether the angle is
a *right angle, less than a right angle,* or *greater than a right angle.*

1.

___**less than a**___
___**right angle**___

2.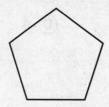

_____

3.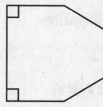

_____

Write how many of each type of angle the shape has.

4.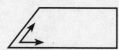

_____ right

_____ less than a right

_____ greater than
a right

5.

_____ right

_____ less than a right

_____ greater than
a right

6.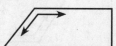

_____ right

_____ less than a right

_____ greater than
a right

Problem Solving REAL WORLD

7. Jeff has a square piece of art paper.
He cuts across it from one corner
to the opposite corner to make
two pieces. What is the total number
of sides and angles in both of the
new shapes?

8. Kaylee tells Aimee that the shape
of a stop sign has at least one right
angle. Aimee says that there are no
right angles. Who is correct? **Explain.**

_____

_____

_____

_____

© Houghton Mifflin Harcourt Publishing Company

Lesson Check

1. What describes this angle?

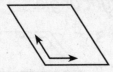

- Ⓐ right angle
- Ⓑ less than a right angle
- Ⓒ greater than a right angle
- Ⓓ small angle

2. How many right angles does this shape have?

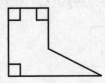

- Ⓐ 1
- Ⓑ 2
- Ⓒ 3
- Ⓓ 4

Spiral Review

3. What fraction of the group is shaded? (Lesson 8.7)

- Ⓐ $\frac{5}{6}$
- Ⓒ $\frac{1}{6}$
- Ⓑ $\frac{1}{3}$
- Ⓓ $\frac{1}{8}$

4. Compare. (Lesson 9.2)

$$\frac{4}{8} \bigcirc \frac{3}{8}$$

- Ⓐ >
- Ⓑ <
- Ⓒ =
- Ⓓ ÷

5. Which of the following does NOT describe a line segment? (Lesson 12.1)

- Ⓐ does not end
- Ⓑ is straight
- Ⓒ is part of a line
- Ⓓ has 2 endpoints

6. How many line segments does this shape have? (Lesson 12.1)

- Ⓐ 5
- Ⓒ 7
- Ⓑ 6
- Ⓓ 8

© Houghton Mifflin Harcourt Publishing Company

Name _____

Identify Polygons

Is the shape a polygon? Write *yes* or *no*.

1.

   __no__

2.

   _____

3.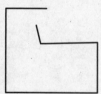

   _____

4.

   _____

Write the number of sides and the number of angles. Then name the polygon.

5.

   _____ sides

   _____ angles

   _____

6.

   _____ sides

   _____ angles

   _____

Problem Solving REAL WORLD

7. Mr. Murphy has an old coin that has ten sides. If its shape is a polygon, how many angles does the old coin have?

   _____

8. Lin says that an octagon has six sides. Chris says that it has eight sides. Whose statement is correct?

   _____

© Houghton Mifflin Harcourt Publishing Company

Lesson Check

1. Which is a name for this polygon?

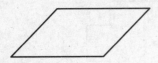

(A) hexagon

(B) octagon

(C) quadrilateral

(D) pentagon

2. How many sides does this polygon have?

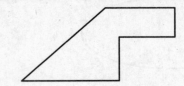

(A) 4

(B) 5

(C) 6

(D) 7

Spiral Review

3. How many right angles does this shape have? (Lesson 12.2)

(A) 4   (C) 2

(B) 3   (D) 0

4. Erica has 8 necklaces. One fourth of the necklaces are blue. How many necklaces are blue? (Lesson 8.9)

(A) 2

(B) 3

(C) 4

(D) 8

5. Which of these is straight, is part of a line, and has 2 endpoints?

(Lesson 12.1)

(A) line

(B) line segment

(C) point

(D) ray

6. What describes this angle? (Lesson 12.2)

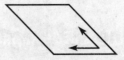

(A) greater than a right angle

(B) large angle

(C) less than a right angle

(D) right angle

© Houghton Mifflin Harcourt Publishing Company

Name _____

Describe Sides of Polygons

Look at the dashed sides of the polygon. Tell if they appear to be *intersecting, perpendicular,* or *parallel.*
Write all the words that describe the sides.

1.

_____ parallel _____

2.

_____

3.

_____

4.

_____

5.

_____

6.

_____

7.

_____

8.

_____

9.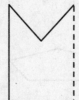

_____

Problem Solving REAL WORLD

Use shapes *A–D* for 10–11.

10. Which shapes appear to have parallel sides?

_____

11. Which shapes appear to have perpendicular sides?

_____

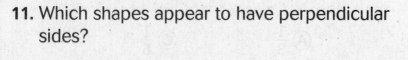

© Houghton Mifflin Harcourt Publishing Company

Lesson Check

1. How many pairs of parallel sides does the quadrilateral appear to have?

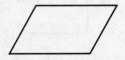

Ⓐ 1   Ⓒ 3

Ⓑ 2   Ⓓ 4

2. Which sides appear to be parallel?

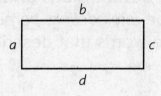

Ⓐ *a* and *c* only

Ⓑ *b* and *d* only

Ⓒ *a* and *b*, *c* and *d*

Ⓓ *a* and *c*, *b* and *d*

Spiral Review

3. Mr. Lance designed a class banner shaped like the polygon shown. What is the name of the polygon?

(Lesson 12.3)

Ⓐ pentagon   Ⓒ hexagon

Ⓑ octagon   Ⓓ decagon

4. How many angles greater than a right angle does this shape have?

(Lesson 12.2)

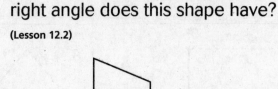

Ⓐ 0   Ⓒ 2

Ⓑ 1   Ⓓ 3

5. How many line segments does this shape have? **(Lesson 12.1)**

Ⓐ 6   Ⓒ 8

Ⓑ 7   Ⓓ 9

6. Which fraction names the shaded part? **(Lesson 8.3)**

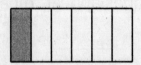

Ⓐ $\frac{1}{3}$ Ⓒ $\frac{1}{6}$

Ⓑ $\frac{1}{4}$ Ⓓ $\frac{5}{6}$

© Houghton Mifflin Harcourt Publishing Company

Classify Quadrilaterals

Circle all the words that describe the quadrilateral.

1.

   (square)

   (rectangle)

   (rhombus)

   trapezoid

2.

   square

   rectangle

   rhombus

   trapezoid

3.

   square

   rectangle

   rhombus

   trapezoid

Use the quadrilaterals below for 4–6.

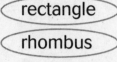 A   B    C   D    E

4. Which quadrilaterals appear to have no right angles?

   _____

5. Which quadrilaterals appear to have 4 right angles?

   _____

6. Which quadrilaterals appear to have 4 sides of equal length?

   _____

Problem Solving REAL WORLD

7. A picture on the wall in Jeremy's classroom has 4 right angles, 4 sides of equal length, and 2 pairs of opposite sides that are parallel. What quadrilateral best describes the picture?

   _____

8. Sofia has a plate that has 4 sides of equal length, 2 pairs of opposite sides that are parallel, and no right angles. What quadrilateral best describes the plate?

   _____

© Houghton Mifflin Harcourt Publishing Company

Lesson Check

1. Which word describes the quadrilateral?

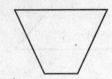

- Ⓐ square
- Ⓑ trapezoid
- Ⓒ rhombus
- Ⓓ rectangle

2. Which quadrilaterals appear to have 2 pairs of opposite sides that are parallel?

- Ⓐ A and B
- Ⓑ A, B, and C
- Ⓒ A only
- Ⓓ B only

Spiral Review

3. Aiden drew the the polygon shown. What is the name of the polygon he drew? (Lesson 12.3)

- Ⓐ decagon
- Ⓒ octagon
- Ⓑ hexagon
- Ⓓ pentagon

4. How many pairs of parallel sides does this shape appear to have? (Lesson 12.4)

- Ⓐ 4
- Ⓒ 1
- Ⓑ 2
- Ⓓ 0

5. What word describes the dashed sides of the shape shown? (Lesson 12.4)

- Ⓐ intersecting
- Ⓒ perpendicular
- Ⓑ parallel
- Ⓓ right

6. How many right angles does this shape have? (Lesson 12.2)

- Ⓐ 0
- Ⓒ 2
- Ⓑ 1
- Ⓓ 3

© Houghton Mifflin Harcourt Publishing Company

Name _____

Draw Quadrilaterals

**Draw a quadrilateral that is described.
Name the quadrilateral you drew.**

1. 4 sides of equal length

square
_____

2. 1 pair of opposite sides that
are parallel

_____

**Draw a quadrilateral that does not belong.
Then explain why.**

3.

_____

_____

_____

Problem Solving REAL WORLD

4. Layla drew a quadrilateral with
4 right angles and 2 pairs of
opposite sides that are parallel.
Name the quadrilateral she could
have drawn.

5. Victor drew a quadrilateral with no
right angles and 4 sides of equal
length. What quadrilateral could
Victor have drawn?

_____

_____

© Houghton Mifflin Harcourt Publishing Company

Lesson Check

1. Chloe drew a quadrilateral with 2 pairs of opposite sides that are parallel. Which shape could NOT be Chloe's quadrilateral?

Ⓐ rectangle

Ⓑ rhombus

Ⓒ square

Ⓓ trapezoid

2. Mike drew a quadrilateral with four right angles. Which shape could he have drawn?

Ⓐ rectangle

Ⓑ hexagon

Ⓒ trapezoid

Ⓓ triangle

Spiral Review

3. A quadrilateral has 4 right angles and 4 sides of equal length. What is the name of the quadrilateral? (Lesson 12.5)

Ⓐ pentagon

Ⓑ square

Ⓒ trapezoid

Ⓓ hexagon

4. Mark drew two lines that form a right angle. Which word describes the lines Mark drew? (Lesson 12.4)

Ⓐ perpendicular

Ⓑ parallel

Ⓒ acute

Ⓓ obtuse

5. Dennis drew the rectangle on grid paper. What is the perimeter of the rectangle Dennis drew? (Lesson 11.2)

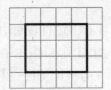

Ⓐ 7 units

Ⓑ 12 units

Ⓒ 14 units

Ⓓ 15 units

6. Jill drew the rectangle on grid paper. What is the area of the rectangle Jill drew? (Lesson 11.5)

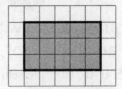

Ⓐ 12 square units

Ⓑ 15 square units

Ⓒ 16 square units

Ⓓ 18 square units

© Houghton Mifflin Harcourt Publishing Company

Name _____

Describe Triangles

Use the triangles for 1–3. Write *A*, *B*, or *C*.
Then complete the sentences.

1. Triangle __*B*__ has 3 angles less than a right angle and

   appears to have __3__ sides of equal length.

2. Triangle _____ has 1 right angle and appears to have

   _____ sides of equal length.

3. Triangle _____ has 1 angle greater than a right angle

   and appears to have _____ sides of equal length.

4. Kyle, Kathy, and Kelly each drew a
   triangle. Who drew the triangle that
   has 1 angle greater than a right angle
   and appears to have no sides of equal
   length?

   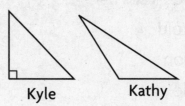

   Kyle   Kathy   Kelly

   _____

Problem Solving REAL WORLD

5. Matthew drew the back of his tent.
   How many sides appear to be of
   equal length?

   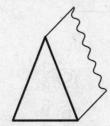

   _____

6. Sierra made the triangular picture
   frame shown. How many angles are
   greater than a right angle?

   _____

© Houghton Mifflin Harcourt Publishing Company

Lesson Check

1. How many angles less than a right angle does this triangle have?

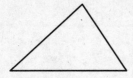

   Ⓐ 0   Ⓒ 2
   Ⓑ 1   Ⓓ 3

2. How many sides of equal length does this triangle appear to have?

   Ⓐ 0   Ⓒ 2
   Ⓑ 1   Ⓓ 3

Spiral Review

3. A quadrilateral has 4 right angles and 2 pairs of opposite sides that are parallel. Which quadrilateral could it be? (Lesson 12.5)

   Ⓐ trapezoid

   Ⓑ hexagon

   Ⓒ triangle

   Ⓓ rectangle

4. Mason drew a quadrilateral with only one pair of opposite sides that are parallel. Which quadrilateral did Mason draw? (Lesson 12.6)

   Ⓐ square

   Ⓑ rhombus

   Ⓒ trapezoid

   Ⓓ rectangle

5. Which shape has an area of 8 square units and a perimeter of 12 units? (Lesson 11.10)

   Ⓐ   Ⓒ

   Ⓑ    Ⓓ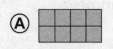

6. What fraction of the square is shaded? (Lesson 8.4)

   Ⓐ $\frac{3}{5}$ Ⓒ $\frac{3}{8}$

   Ⓑ $\frac{5}{3}$ Ⓓ $\frac{8}{3}$

© Houghton Mifflin Harcourt Publishing Company

Name _____

Problem Solving • Classify Plane Shapes

Solve each problem.

1. Steve drew the shapes below. Write the letter of each shape where it belongs in the Venn diagram.

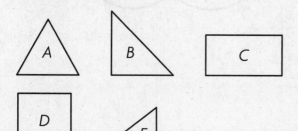

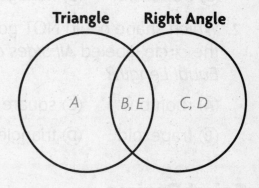

Triangle   Right Angle

A   B, E   C, D

2. Janice drew the shapes below. Write the letter of each shape where it belongs in the Venn diagram.

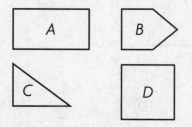

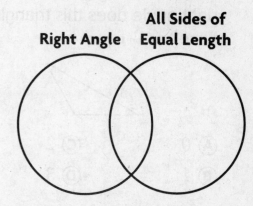

Right Angle   All Sides of Equal Length

3. Beth drew the shapes below. Write the letter of each shape where it belongs in the Venn diagram.

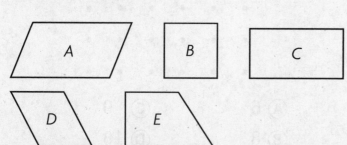

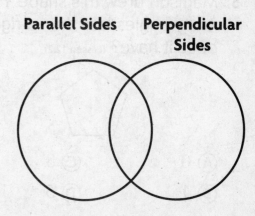

Parallel Sides   Perpendicular Sides

© Houghton Mifflin Harcourt Publishing Company

Lesson Check

1. Which shape would go in the section where the two circles overlap?

   Ⓐ triangle   Ⓒ square

   Ⓑ trapezoid   Ⓓ hexagon

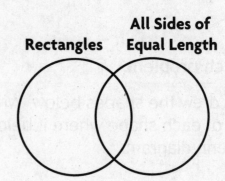

Rectangles   **All Sides of Equal Length**

2. Which shape could NOT go in the circle labeled *All Sides of Equal Length*?

   Ⓐ rhombus   Ⓒ square

   Ⓑ trapezoid   Ⓓ triangle

Spiral Review

3. How many angles greater than a right angle does this triangle have? (Lesson 12.7)

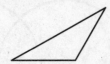

   Ⓐ 0   Ⓒ 2

   Ⓑ 1   Ⓓ 3

4. How many sides of equal length does this triangle appear to have? (Lesson 12.7)

   Ⓐ 0   Ⓒ 2

   Ⓑ 1   Ⓓ 3

5. Madison drew this shape. How many angles less than a right angle does it have? (Lesson 12.2)

   Ⓐ 0   Ⓒ 3

   Ⓑ 1   Ⓓ 5

6. How many dots are in $\frac{1}{2}$ of this group? (Lesson 8.7)

   Ⓐ 6   Ⓒ 9

   Ⓑ 8   Ⓓ 18

© Houghton Mifflin Harcourt Publishing Company

Name _____

Relate Shapes, Fractions, and Area

Draw lines to divide the shape into equal parts that show the fraction given.

1.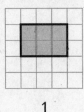

$\frac{1}{3}$

2.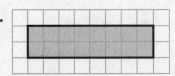

$\frac{1}{8}$

3.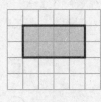

$\frac{1}{2}$

Draw lines to divide the shape into parts with equal area. Write the area of each part as a unit fraction.

4.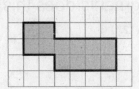

4 equal parts

_____

5.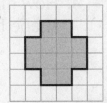

6 equal parts

_____

6.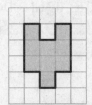

3 equal parts

_____

Problem Solving REAL WORLD

7. Robert divided a hexagon into 3 equal parts. Show how he might have divided the hexagon. Write the fraction that names each part of the whole you divided.

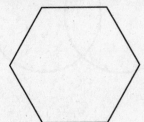

8. Show how you might divide the shape into 8 equal parts. What fraction names the area of each part of the divided shape?

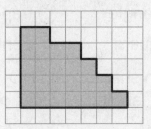

_____   _____

© Houghton Mifflin Harcourt Publishing Company

Lesson Check

1. What fraction names each part of the divided whole?

   Ⓐ $\frac{1}{2}$ Ⓒ $\frac{1}{4}$

   Ⓑ $\frac{1}{3}$ Ⓓ $\frac{1}{6}$

2. What fraction names the whole area that was divided?

   Ⓐ $\frac{1}{8}$ Ⓒ $\frac{8}{8}$

   Ⓑ $\frac{1}{2}$ Ⓓ $\frac{8}{1}$

Spiral Review

3. Lil drew the figure below. Which word does NOT describe the shape? (Lesson 12.1)

   Ⓐ plane shape

   Ⓑ closed shape

   Ⓒ open shape

   Ⓓ curved path

4. How many line segments does this shape have? (Lesson 12.1)

   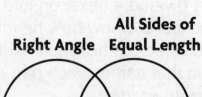

   Ⓐ 6

   Ⓑ 5

   Ⓒ 4

   Ⓓ 3

Use the Venn diagram for 5–6. (Lesson 12.8)

5. Which shape would go in the section where the two circles overlap?

   Ⓐ triangle   Ⓒ trapezoid

   Ⓑ square   Ⓓ pentagon

6. Which shape could NOT go in the circle labeled *All Sides of Equal Length*?

   Ⓐ square   Ⓒ triangle

   Ⓑ rhombus   Ⓓ rectangle

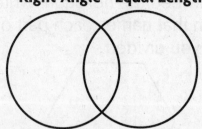

Right Angle   All Sides of Equal Length

© Houghton Mifflin Harcourt Publishing Company

Name _____

Chapter 12 Extra Practice

Lessons 12.1 - 12.3

Name the polygon.

1.

_____

2.

_____

3.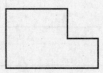

_____

4.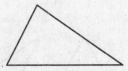

_____

5.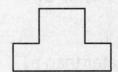

_____

6.

_____

Lesson 12.4

Look at the dashed sides of the polygon. Tell if they appear to be *intersecting*, *perpendicular*, or *parallel*. Write all the words that describe the sides.

1.

_____

2.

_____

3.

_____

Lesson 12.5

Circle all the words that describe the quadrilateral.

1.

rhombus

trapezoid

rectangle

2.

square

rhombus

trapezoid

3.

trapezoid

rectangle

rhombus

© Houghton Mifflin Harcourt Publishing Company

Lesson 12.6

Draw a quadrilateral that does not belong. Then explain why.

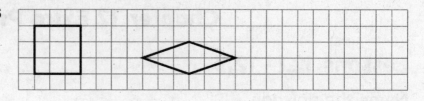

_____

_____

Lesson 12.7

Use the triangles for 1–2. Write *A*, *B*, or *C*. Then complete the sentences.

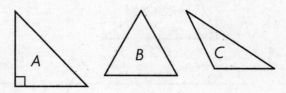

1. Triangle _____ has 1 angle greater than a right angle and appears to have _____ sides of equal length.

2. Triangle _____ has 1 right angle and appears to have _____ sides of equal length.

Lesson 12.8

1. What label could you use to describe Circle *A*?

_____

_____

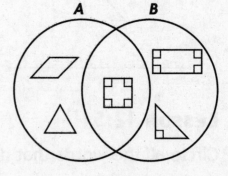

2. What label could you use to describe Circle *B*?

_____

_____

Lesson 12.9

Draw lines to divide the shape into equal parts that show the fraction given.

1. $\frac{1}{4}$

2. $\frac{1}{3}$

© Houghton Mifflin Harcourt Publishing Company

Name _____

Numbers to Ten Thousand

Essential Question How can you represent numbers to ten thousand in different ways?

🔓 UNLOCK the Problem REAL WORLD

The Thousand Bolts factory uses boxes of 1,000 bolts to fill crates of 10,000 bolts. How many boxes of 1,000 bolts are in each crate of 10,000?

> • Circle the number you will need to count to find the answer.

 Count by thousands to find the total number of boxes of 1,000 bolts that will go into each crate. Then count the boxes.

1,000 **2,000** _____ _____ _____ _____ _____
[1] [2] [ ] [ ] [ ] [ ] [ ]

_____ _____ _____
[ ] [ ] [ ]

So, there are _____ boxes of 1,000 bolts in each crate of 10,000.

🔑 Example
Suppose the factory has no crates and must use cases of 100 to fill an order for 3,200 bolts. How many cases will it pack?

There are _____ cases of 100 in 1,000.

So, there are _____ cases of 100 in 3,000.

There are _____ cases of 100 in 200.

Add the cases. 30 + 2 = _____.

So, the factory will pack 32 cases of 100.

> **Math Talk** What if the factory had boxes of 1,000 and bags of 10 but no cases of 100? **Explain** how it could pack the order for 3,200 bolts.

© Houghton Mifflin Harcourt Publishing Company

Share and Show

1. The Thousand Bolts factory has an order for 3,140 bolts. How can it pack the order using the fewest packages?

_____

Remember

1 box = 1,000 bolts
1 case = 100 bolts
1 bag = 10 bolts

2. Suppose the bolt factory has only cases and bags. How can it pack the order for 3,140 bolts?

_____

3. Suppose the bolt factory has only boxes and bags. How can it pack the order for 3,140 bolts?

_____

On Your Own

Complete the packing chart. Use the fewest packages possible. When there is a zero, use the next smaller size package.

	Number of Bolts Ordered	Crates (Ten Thousands)	Boxes (Thousands)	Cases (Hundreds)	Bags (Tens)	Single Bolts (Ones)
4.	5,267		5			
5.	2,709			7	0	
6.	5,619					
7.	8,416		0		1	6
8.	3,967		0		0	

Problem Solving REAL WORLD

9. The Thousand Bolts factory used 9 boxes, 9 cases, and 10 bags to fill an order. How many bolts did they pack?

_____

© Houghton Mifflin Harcourt Publishing Company

Read and Write Numbers to Ten Thousands

Essential Question What are some ways you can read and write numbers?

UNLOCK the Problem REAL WORLD

The ABC Block Factory receives an order for blocks. The base-ten blocks show the number of blocks ordered.

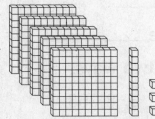

- How many blocks were ordered?
  _____

Math Idea

The location of a digit in a number tells its value.

Each worker on the team checks the order by expressing the number in a different way. What way does each worker use?

 Read and write numbers.

Word form is a way to write a number using words.

Sam gets the order and reads the number to Mary: two thousand, five hundred thirteen

Expanded form is a way to write a number by showing the value of each digit.

Mary uses the value of each digit to record the number of blocks that will be in each type of package:
2,000 + 500 + 10 + 3

Standard form is a way to write a number using the digits 0 to 9, with each digit having a place value.

When the order is complete, Kyle writes the total number of blocks on the packing slip: 2,513

So, Sam says the number using _____

form, Mary uses _____ form,

and Kyle uses _____ form.

Math Talk **Explain** how to find the value of the underlined digit in 7,521.

© Houghton Mifflin Harcourt Publishing Company

Share and Show

1. Write the number shown in expanded form.

TEN THOUSANDS	THOUSANDS	HUNDREDS	TENS	ONES
	7,	5	9	8

_____ + 500 + 90 + _____

Write the number in standard form.

2. 4,000 + 600 + 70 + 4   _____

3. eight thousand, two hundred sixty-one _____

Write the value of the underlined digit two ways.

4. 6,920

5. 8,063

_____

On Your Own ....

Write the number in standard form.

6. 5,000 + 600 + 90 + 7   _____

7. two thousand, three hundred fifty-nine _____

8. one thousand, three hundred two _____

Write the value of the underlined digit two ways.

9. 6,818

10. 9,342

_____   _____

11. Rename 3,290 as hundreds and tens.

12. Rename 2,934 as tens and ones.

_____ hundreds _____ tens

_____ tens _____ ones

Problem Solving REAL WORLD

13. The number of children who attended the fair on opening day is 351 more than the value of 4 thousands. How many children attended the fair on opening day? _____

© Houghton Mifflin Harcourt Publishing Company

Name _____

Relative Size on a Number Line

Essential Question How can you locate and name a point on a number line?

🔓 UNLOCK the Problem REAL WORLD

Wilfren has 40 pennies, Ella has 400 pennies, and Matt has 4,000 pennies. How do their amounts of pennies compare?

> • Circle the amounts you need to compare.

🔑 **Compare the relative sizes of the amounts of pennies.**

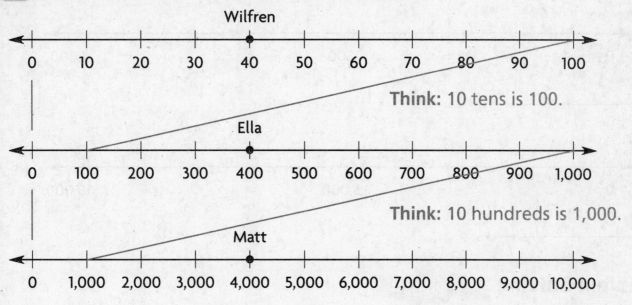

Think: 10 tens is 100.

Think: 10 hundreds is 1,000.

Think: 10 thousands is 10,000.

So, Ella has _____ times as many pennies as

Wilfren, and Matt has _____ times as many pennies as Ella.

Try This! Find the number represented by the point.

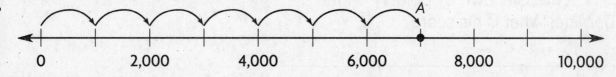

Start at 0. Skip count by 1,000s until you reach point A.

There are _____ jumps of 1,000. So, point A represents

_____ .

> **Math Talk**
> **Explain** how to locate and draw the point 3,000 on a number line.

© Houghton Mifflin Harcourt Publishing Company

Share and Show

Find the number that point *B* represents on the number line.

1.

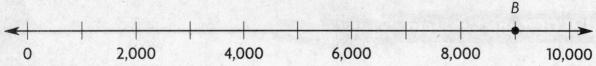

_____

On Your Own

Find the number represented by the point.

2.

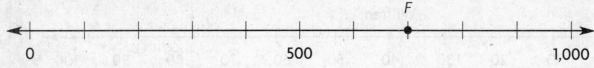

_____

3.

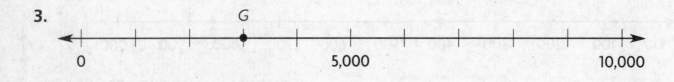

_____

Problem Solving

Use the number line for 4–5.

Nestor and Elliot are playing a number line game.

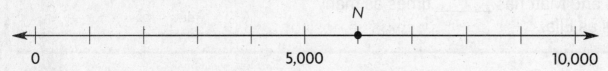

4. Nestor's score is shown by point *N* on the number line. What is his score?

_____

5. Elliot's score is 8,000. Is Elliot's score located to the right or to the left of Nestor's score? **Explain**.

_____

© Houghton Mifflin Harcourt Publishing Company

Name _____

Compare 3- and 4-Digit Numbers

Essential Question What are some ways you can compare numbers?

🔑 UNLOCK the Problem   REAL WORLD

Cody collected 2,365 pennies. Jasmine collected 1,876 pennies. Who collected more pennies?

You can compare numbers in different ways to find which number is greater.

* What do you need to find?

_____

_____

🔑 One Way Use base-ten blocks.

Compare the values of the blocks in each place-value position from left to right. Keep comparing the blocks until the values are different.

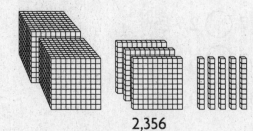

2,356

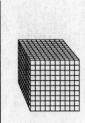
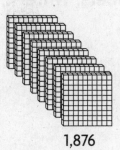

1,876

2 thousands is greater than 1 thousand. So, 2,365 ◯ 1,876.

So, Cody collected more pennies.

🔑 Another Way Use place value.

Compare 7,376 and 7,513.

Compare digits in the same place-value position from left to right.

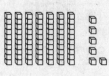

Read Math

Read < as *is less than*.
Read > as *is greater than*.
Read = as *is equal to*.

THOUSANDS	HUNDREDS	TENS	ONES
7,	3	7	6
7,	5	1	3

STEP 1: Compare the thousands. The digits are the same.

STEP 2: Compare the hundreds. 3 ◯ 5

So, 7,376 ◯ 7,513.

Math Talk **Explain** how you know that 568 is less than 4,786.

© Houghton Mifflin Harcourt Publishing Company

Share and Show

1. Compare 2,351 and 3,018. Which number has more thousands? Which number is greater?

_____

_____

Compare the numbers. Write <, >, or = in the ◯.

2. 835 ◯ 853

3. 7,891 ◯ 7,891

4. 809 ◯ 890

5. 3,834 ◯ 3,483

On Your Own

Compare the numbers. Write <, >, or = in the ◯.

6. 219 ◯ 2,119

7. 2,517 ◯ 2,715

8. 5,154 ◯ 5,154

9. 5,107 ◯ 5,105

10. 1,837 ◯ 837

11. 9,832 ◯ 9,328

Problem Solving REAL WORLD

12. Nina has a dictionary with 1,680 pages. Trey has a dictionary with 1,490 pages. Use <, >, or = to compare the number of pages in the dictionaries.

_____

13. The odometer in Ed's car shows it has been driven 8,946 miles. The odometer in Beth's car shows it has been driven 5,042 miles. Which car has been driven more miles?

_____

14. Avery said that she is 3,652 days old. Tamika said that she is 3,377 days old. Who is younger?

_____

© Houghton Mifflin Harcourt Publishing Company

Name _____

Concepts and Skills

Complete the packing chart. Use the fewest packages
possible. When there is a zero, use the next smaller
size package. (pp. P259–P260)

	Number of Bolts Ordered	Crates (Ten Thousands)	Boxes (Thousands)	Cases (Hundreds)	Bags (Tens)	Single Bolts (Ones)
1.	5,267		5			
2.	2,709			7	0	

**Find the number that point *A* represents on the
number line.** (pp. P263–P264)

3.
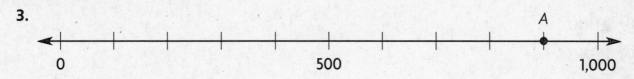

_____

Compare the numbers. Write <, >, or = in the ◯. (pp. P265–P266)

4. 4,310 ◯ 4,023

5. 5,136 ◯ 5,136

6. 732 ◯ 6,532

7. 9,436 ◯ 4,963

Problem Solving  (pp. P261–P262, P265–P266)

8. The number of people who attended
the Spring Festival is 799 more than
8 thousands. How many people
attended the festival?

9. There are 1,290 photos on
Nadia's memory card. There are
1,450 photos on Trevor's memory
card. Use <, >, or = to compare
the number of photos on the
memory cards.

_____   _____

© Houghton Mifflin Harcourt Publishing Company

Fill in the bubble for the correct answer choice.

10. A marble factory ships marbles using bags of 10, cases of 100, cartons of 1,000, and boxes of 10,000. The factory has an order for 3,570 marbles. How can they pack the order if the factory is out of cartons? (pp. P259–P260)

Ⓐ 350 cases, 7 bags

Ⓑ 35 cases, 7 bags

Ⓒ 35 cases, 57 bags

Ⓓ 3 cases, 75 bags

11. The number of fans who attend the baseball game on opening day is 283 more than 4 thousands. How many fans are attending the baseball game on opening day? (pp. P261–P262)

Ⓐ 283

Ⓑ 4,000

Ⓒ 4,283

Ⓓ 4,823

Use the number line for 12–13.

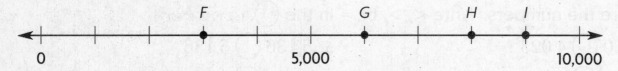

12. Kam scored 6,000 points in a game. Which letter on the number line names the point that represents Kam's score? (pp. P263–P264)

Ⓐ F 　　Ⓒ H

Ⓑ G 　　Ⓓ I

13. Taissa scored 9,000 points in a game. Which letter on the number line names the point that represents Taissa's score? (pp. P263–P264)

Ⓐ F 　　Ⓒ H

Ⓑ G 　　Ⓓ I

© Houghton Mifflin Harcourt Publishing Company

Name _____

Multiply with 11 and 12

Essential Question What strategies can you use to multiply with 11 and 12?

🔓 UNLOCK the Problem · REAL WORLD

It takes Bobby 11 minutes to walk to school each morning. How many minutes will Bobby spend walking to school in 5 days?

- What are the groups in this problem?

  _____

  _____

Multiply. $5 \times 11 = $ ▢

🔑 One Way Break apart an array.

Make 5 rows of 11. Use the 10s facts and the 1s facts to multiply with 11.

$5 \times (10 + 1)$

$5 \times 10 = $ _____ $5 \times 1 = $ _____

$5 \times 11 = $ _____ + _____

$5 \times 11 = $ _____

🔑 Another Way Find a pattern.

Look at the list.

Notice the product has the same factor in the tens and ones places.

To find 5×11, write the first factor in the tens and ones places.

$5 \times 11 = 55$

$1 \times 11 = \ 11$
$2 \times 11 = \ 22$
$3 \times 11 = \ 33$
$4 \times 11 = \ 44$

$5 \times 11 = \ $ _____
$6 \times 11 = \ 66$
$7 \times 11 = \ 77$
$8 \times 11 = \ 88$
$9 \times 11 = \ 99$

So, Bobby will spend _____ minutes walking to school.

Try This! What if it took Bobby 12 minutes to walk to school? How many minutes will he spend walking to school in 5 days?

Break apart the factor 12.

$5 \times (10 + 2)$

$5 \times 10 = 50$ $5 \times 2 = 10$

$5 \times 12 = $ _____ + _____ = _____

Double a 6s fact.

Find the 6s product. $5 \times 6 = 30$

Double that product. _____ + _____ = _____

So, $5 \times 12 = $ _____. Bobby will spend _____ minutes walking to school.

© Houghton Mifflin Harcourt Publishing Company

Share and Show

1. How can you use the 10s facts and the 2s facts to find 4×12?

_____

_____

Find the product.

2. $9 \times 11 = $ _____ | 3. $7 \times 12 = $ _____ | 4. _____ $= 4 \times 11$

On Your Own ....

Find the product.

5. _____ $= 11 \times 6$ | 6. _____ $= 12 \times 2$ | 7. $0 \times 11 = $ _____

8. _____ $= 6 \times 12$ | 9. $8 \times 12 = $ _____ | 10. $7 \times 11 = $ _____

11. $12 \times 9 = $ _____ | 12. $3 \times 12 = $ _____ | 13. $1 \times 12 = $ _____

Problem Solving REAL WORLD

Use the graph for 14–15.

14. The graph shows the number of miles some students travel to school each day. How many miles will Carlos travel to school in 5 days?

_____

15. Suppose that Mandy takes 9 trips to school, and Matt takes 11 trips to school. Who travels more miles? **Explain.**

_____

_____

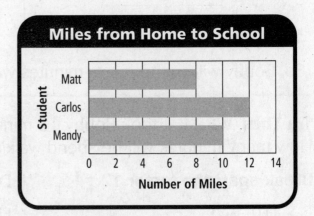

Miles from Home to School

© Houghton Mifflin Harcourt Publishing Company

Divide with 11 and 12

Essential Question What strategies can you use to divide with 11 and 12?

🔓 UNLOCK the Problem REAL WORLD

Tara collects 60 postcards. She arranges them in 12 equal stacks. How many postcards are in each stack?

Divide. $60 \div 12 = \blacksquare$

- Do you need to find the number of groups or the number in each group?

_____

_____

🔑 One Way Use a multiplication table.

Since division is the inverse of multiplication, you can use a multiplication table to find a quotient.

Think of a related multiplication fact.

$$12 \times \blacksquare = 60$$

- Find the row for the factor 12.
- Look across to find the product, 60.
- Look up to find the unknown factor.
- The unknown factor is 5.

Since $12 \times 5 = 60$, then

$60 \div 12 = $ _____.

×	0	1	2	3	4	5	6	7	8	9	10	11	12
0	0	0	0	0	0	0	0	0	0	0	0	0	0
1	0	1	2	3	4	5	6	7	8	9	10	11	12
2	0	2	4	6	8	10	12	14	16	18	20	22	24
3	0	3	6	9	12	15	18	21	24	27	30	33	36
4	0	4	8	12	16	20	24	28	32	36	40	44	48
5	0	5	10	15	20	25	30	35	40	45	50	55	60
6	0	6	12	18	24	30	36	42	48	54	60	66	72
7	0	7	14	21	28	35	42	49	56	63	70	77	84
8	0	8	16	24	32	40	48	56	64	72	80	88	96
9	0	9	18	27	36	45	54	63	72	81	90	99	108
10	0	10	20	30	40	50	60	70	80	90	100	110	120
11	0	11	22	33	44	55	66	77	88	99	110	121	132
12	0	12	24	36	48	60	72	84	96	108	120	132	144

🔑 Another Way Use repeated subtraction.

- Start with 60.
- Subtract 12 until you reach 0.
- Count the number of times you subtract 12.

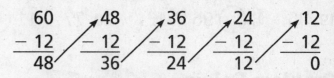

You subtracted 12 five times.

$60 \div 12 = $ _____

So, there are 5 postcards in each stack.

Math Talk What other strategies can you use to divide?

© Houghton Mifflin Harcourt Publishing Company

Share and Show

1. Use the multiplication table on page P271 to find 99 ÷ 11.

                            Think: What is a related multiplication fact?

_____

Find the unknown factor and quotient.

2. $11 \times \blacksquare = 66$ $66 \div 11 = \blacksquare$ **3.** $2 \times \blacksquare = 24$ $24 \div 2 = \blacksquare$

    $\blacksquare = $ _____ $\blacksquare = $ _____ $\blacksquare = $ _____ $\blacksquare = $ _____

4. $3 \times \blacksquare = 33$ $33 \div 3 = \blacksquare$ **5.** $12 \times \blacksquare = 72$ $72 \div 12 = \blacksquare$

    $\blacksquare = $ _____ $\blacksquare = $ _____ $\blacksquare = $ _____ $\blacksquare = $ _____

On Your Own . . . .

Find the unknown factor and quotient.

6. $11 \times \blacksquare = 55$ $55 \div 11 = \blacksquare$ **7.** $12 \times \blacksquare = 48$ $48 \div 12 = \blacksquare$

    $\blacksquare = $ _____ $\blacksquare = $ _____ $\blacksquare = $ _____ $\blacksquare = $ _____

8. $8 \times \blacksquare = 96$ $96 \div 8 = \blacksquare$ **9.** $8 \times \blacksquare = 88$ $88 \div 8 = \blacksquare$

    $\blacksquare = $ _____ $\blacksquare = $ _____ $\blacksquare = $ _____ $\blacksquare = $ _____

Find the quotient.

10. $11 \div 11 = $ _____   **11.** $77 \div 7 = $ _____   **12.** _____ $= 60 \div 12$

13. _____ $= 22 \div 11$ **14.** $108 \div 9 = $ _____   **15.** $84 \div 12 = $ _____

16. $36 \div 3 = $ _____   **17.** _____ $= 96 \div 12$ **18.** $12 \div 12 = $ _____

Compare. Write <, >, or = for each $\bigcirc$.

19. $96 \div 8 \bigcirc 96 \div 12$ **20.** $77 \div 11 \bigcirc 84 \div 12$ **21.** $99 \div 11 \bigcirc 84 \div 7$

Problem Solving REAL WORLD

22. Justin printed 44 posters to advertise the garage sale. He gave 11 friends the same number of posters to display around the neighborhood. How many posters did Justin give each friend?

_____

© Houghton Mifflin Harcourt Publishing Company

Name _____

Multiplication and Division Relationships

Essential Question How can you write related multiplication and division equations for 2-digit factors?

Multiplication and division are inverse operations.

🔓 UNLOCK the Problem REAL WORLD

Megan has a rose garden with the same number of bushes planted in each of 4 rows. There are 48 bushes in the garden. How many bushes are in each row of Megan's garden?

- What do you need to find?

_____

_____

🔑 One Way

Make an array.

48 ÷ 4 = ■

Count 48 tiles. Make 4 rows by placing 1 tile in each row.

Continue placing 1 tile in each of the 4 rows until all the tiles are used.

Draw the array you made.

There are _____ tiles in each row.

_____ ÷ _____ = _____

So, there are _____ bushes in each row of Megan's garden.

🔑 Another Way

Write related equations.

48 ÷ 4 = ■

Think: 4 times what number equals 48?

4 × _____ = 48

You can check your answer using repeated addition.

_____ + _____ + _____ + _____ = _____

Write related equations.

_____ × _____ = 48

48 ÷ _____ = _____

Math Talk How can you tell if two equations are related?

© Houghton Mifflin Harcourt Publishing Company

Share and Show

1. Complete the related equations for this array.

$3 \times 11 = 33$ $33 \div 3 = 11$

_____   _____

Complete the related multiplication and division equations.

2. $1 \times 11 = $ _____

   _____ $\times 1 = 11$

   $11 \div 1 = $ _____

   _____ $\div 11 = 1$

3. $5 \times $ _____ $= 60$

   $12 \times 5 = $ _____

   _____ $\div 5 = 12$

   $60 \div $ _____ $= 5$

4. _____ $\times 11 = 77$

   _____ $\times 7 = 77$

   $77 \div $ _____ $= 11$

   _____ $\div 11 = 7$

On Your Own

Complete the related multiplication and division equations.

5. _____ $\times 12 = 84$

   _____ $\times 7 = 84$

   _____ $\div 7 = 12$

   $84 \div $ _____ $= 7$

6. $6 \times $ _____ $= 66$

   $11 \times $ _____ $= 66$

   $66 \div 6 = $ _____

   $66 \div 11 = $ _____

7. $12 \times 8 = $ _____

   $8 \times $ _____ $= 96$

   $96 \div $ _____ $= 8$

   $96 \div 8 = $ _____

Problem Solving

8. Megan cut 108 roses to make flower arrangements. She made 9 equal arrangements. How many roses were in each arrangement?

_____

9. Megan put 22 roses in a vase. She cut the same number of roses from each of 11 different bushes. How many roses did she cut from each bush?

_____

© Houghton Mifflin Harcourt Publishing Company

Name _____

Use Multiplication Patterns

Essential Question How can you multiply with 10, 100, and 1,000?

UNLOCK the Problem

Mrs. Goldman ordered 4 boxes of yo-yos for her toy store. Each box had 100 yo-yos. How many yo-yos did Mrs. Goldman order?

- Circle the numbers you need to use.
- What operation can you use to find the total when you have equal groups?

_____

🔑 Use a basic fact and a pattern to multiply.

Factors		Products
4×1	$=$	4
4×10	$=$	40
4×100	$=$	400

Think: Use the basic fact $4 \times 1 = 4$. Look for a pattern of zeros.

So, Mrs. Goldman ordered 400 yo-yos.

Math Idea

As the number of zeros in a factor increases, the number of zeros in the product increases.

Try This! Use a basic fact and a pattern to find the products.

A. $1 \times 3 = 3$

$10 \times 3 = $ _____

B. $5 \times 1 \quad = 5$

$5 \times 10 \quad = 50$

$5 \times 100 \quad = $ _____

$5 \times 1{,}000 = $ _____

Math Talk When multiplying $9 \times 1{,}000$, how many zeros will be in the product? **Explain.**

© Houghton Mifflin Harcourt Publishing Company

Share and Show

1. **Explain** how to use a basic fact and a pattern to find 6×100.

_____

_____

Use a basic fact and a pattern to find the products.

2. $7 \times 10 =$ _____

 $7 \times 100 =$ _____

 $7 \times 1{,}000 =$ _____

3. $10 \times 5 =$ _____

 $100 \times 5 =$ _____

 $1{,}000 \times 5 =$ _____

4. $3 \times 10 =$ _____

 $3 \times 100 =$ _____

 $3 \times 1{,}000 =$ _____

On Your Own

Use a basic fact and a pattern to find the products.

5. $2 \times 10 =$ _____

 $2 \times 100 =$ _____

 $2 \times 1{,}000 =$ _____

6. $10 \times 8 =$ _____

 $100 \times 8 =$ _____

 $1{,}000 \times 8 =$ _____

7. $9 \times 10 =$ _____

 $9 \times 100 =$ _____

 $9 \times 1{,}000 =$ _____

Find the product.

8. $10 \times 8 =$ _____

9. $6 \times 100 =$ _____

10. _____ $= 4 \times 100$

11. $1{,}000 \times 4 =$ _____

12. _____ $= 1{,}000 \times 3$

13. $9 \times 100 =$ _____

Problem Solving

Use the picture graph.

14. Patty has 20 fewer yo-yos in her collection than Chuck. Draw yo-yos in the picture graph. to show the number of yo-yos in Patty's collection. **Explain** your answer.

_____

_____

Yo-Yo Collections	
Name	**Number of Yo-Yos**
Max	🪀 🪀 🪀
Chuck	🪀 🪀 🪀 🪀
Patty	

Key: Each 🪀 = 10 Yo-Yos.

© Houghton Mifflin Harcourt Publishing Company

Name _____

Use Models to Multiply Tens and Ones

Essential Question How can you use base-ten blocks and area models to model multiplication with a 2-digit factor?

🔓 UNLOCK the Problem REAL WORLD

Three groups of 14 students toured the state capitol in Columbus, Ohio. How many students toured the capitol in all?

Multiply. $3 \times 14 =$ ■

- What do you need to find?

  _____

  _____

- Circle the numbers you need to use.

🔑 One Way

STEP 1

Model 3×14 with base-ten blocks.

3 rows of 10   3 rows of 4

STEP 2

Multiply the tens and ones. Record each product.

$3 \times 10 =$ _____ $3 \times 4 =$ _____

STEP 3

Add the products.
$30 + 12 = 42$

$3 \times 14 = 42$

So, 42 students toured the capitol.

🔑 Another Way

STEP 1

Model 3×14 with an area model.

3 rows of 10   3 rows of 4

STEP 2

Multiply the tens.   Multiply the ones.

$3 \times 10 =$ _____ $3 \times 4 =$ _____

STEP 3

Add the products.
$30 + 12 = 42$

$3 \times 14 = 42$

Math Talk How are the two ways to find a product alike?

© Houghton Mifflin Harcourt Publishing Company

Share and Show

1. One way to model 18 is 1 ten 8 ones. How can knowing this help you find 4×18?

_____

_____

Find the product. Show your multiplication and addition.

2.

$3 \times 16 = \blacksquare$

_____

_____

3.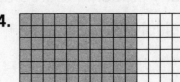

$5 \times 13 = \blacksquare$

_____

_____

4.

$6 \times 14 = \blacksquare$

_____

_____

On Your Own

Find the product. Show your multiplication and addition.

5.

$4 \times 13 = \blacksquare$

_____

_____

6.

$5 \times 15 = \blacksquare$

_____

_____

7.

$3 \times 17 = \blacksquare$

_____

_____

Problem Solving

8. Randy rakes yards for $5 an hour. How much money does he earn if he works for 12 hours? _____

© Houghton Mifflin Harcourt Publishing Company

Name _____

Model Division with Remainders

Essential Question How can you use counters to model division with remainders?

🔒 UNLOCK the Problem REAL WORLD

Madison has 13 seeds. She wants to put the same number of seeds in each of 3 pots. How many seeds can Madison put into each pot? How many seeds are left over?

- How do you know how many groups to make?

  _____

🔑 Activity Materials ▪ counters

Use counters to find 13 ÷ 3.

STEP 1 Use 13 counters. Draw 3 circles for the 3 pots.

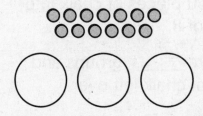

STEP 2 Place one counter in each group until there are not enough to put 1 more in each of the groups.

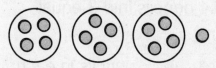

There are _____ counters in each circle.

There is _____ counter left over.

13 ÷ 3 is 4 with 1 left over.

The quotient is 4.

The remainder is 1.

So, Madison can put 4 seeds in each pot. There is 1 seed left over.

After dividing a group of objects into equal groups as large as possible, there may be some left over. The amount left over is called the **remainder**.

Math Talk **Explain** why you cannot have a remainder of 3 when you divide by 3.

Try This! **What if** Madison wants to put 4 seeds in each pot. How many pots will Madison need? How many seeds will be left over?

_____

© Houghton Mifflin Harcourt Publishing Company

Share and Show

1. Divide 13 counters into 2 equal groups.

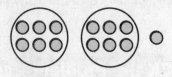

There are _____ counters in each group,

and _____ counter left over.

Complete.

2. April divided 17 counters into 4 equal groups.

There were _____ counters in each

group and _____ counter left over.

3. Divide 20 counters into groups of 6.

There are _____ groups and _____ counters left over.

On Your Own
Complete.

4. Divide 14 pencils into 3 equal groups.

There are _____ pencils in each

group and _____ pencils left over.

5. Divide 60 pieces of chalk into groups of 8.

There are _____ groups and _____ pieces of chalk left over.

Find the total number of objects.

6. There are 2 shoes in each of 6 groups and 1 shoe left over.

There are _____ shoes in all.

7. There are 4 apples in each of 3 groups and 2 apples left over.

There are _____ apples in all.

Problem Solving

Use the bar graph for 8.

8. If Hector divides the oak leaves evenly into 4 display boxes, how many leaves will be in each box? How many leaves will be left over?

_____

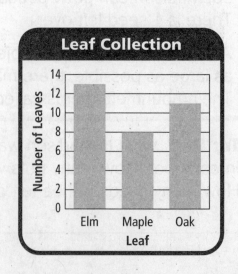

© Houghton Mifflin Harcourt Publishing Company

Name _____

Use Models to Divide Tens and Ones

Essential Question How can you model division with a 2-digit quotient?

🔓 UNLOCK the Problem REAL WORLD

Emma baked 52 muffins. She wants to put an equal number of muffins on each of 4 trays. How many muffins can she put on each tray?

- Circle the numbers you need to use.
- How many equal groups are there?

_____

🔑 Find 52 ÷ 4.

STEP 1

Use base-ten blocks to model the problem. Draw 4 rectangles to represent the 4 equal groups.

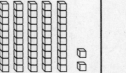
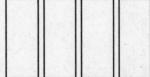

STEP 2

Share the tens. Place 1 ten in each group until there are not enough tens to put 1 more ten in each group.

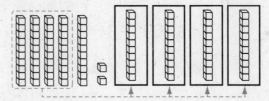

STEP 3

Regroup the remaining ten as ones. There are now 12 ones.

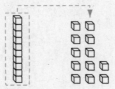

STEP 4

Share the ones. Place 1 one in each group until there are not enough ones to put 1 more one in each group.

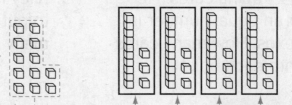

So, Emma can put _____ muffins on each tray.

Math Talk How can you check your answer?

© Houghton Mifflin Harcourt Publishing Company

Share and Show

1. Find $42 \div 2$.

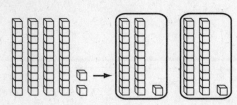

- How many equal groups are there? _____
- How many tens go in each group? _____
- How many ones go in each group? _____
- The quotient is _____.

Use base-ten blocks and your MathBoard to divide.

2. $65 \div 5 =$ _____

3. $90 \div 3 =$ _____

4. $88 \div 4 =$ _____

On Your Own

Use base-ten blocks and your MathBoard to divide.

5. $72 \div 2 =$ _____

6. $69 \div 3 =$ _____

7. $96 \div 6 =$ _____

Problem Solving

8. Roger has 84 trading cards. He wants to put an equal number in each of 3 boxes. How many cards will he put into each box?

_____

9. Riley has 78 postcards. She wants to put 6 on each poster board. How many poster boards will she need?

_____

© Houghton Mifflin Harcourt Publishing Company

Name _____

✓ Checkpoint

Concepts and Skills

Find the product. (pp. P269–P270)

1. _____ = 11 × 5

2. 12 × 7 = _____

Find the unknown factor and quotient. (pp. P271–P272)

3. 4 × ■ = 44   44 ÷ 4 = ■

■ = _____   ■ = _____

4. Write the related multiplication and division equations
   for the numbers 5, 12, 60. (pp. P273–P274)

_____ _____ _____ _____

Use a basic fact and a pattern to find the products. (pp. P275–P276)

5. 3 × 10 = _____

   3 × 100 = _____

   3 × 1,000 = _____

6. 10 × 7 = _____

   100 × 7 = _____

   1,000 × 7 = _____

Find the product. Show your multiplication and division. (pp. P277–P278)

7.

3 × 10 = _____   3 × 4 = _____

3 × 14 = ■

_____ + _____ = _____

3 × 14 = _____

Use base-ten blocks and your MathBoard to divide. (pp. P281–P282)

8. 132 ÷ 6 = _____

9. 160 ÷ 8 = _____

Problem Solving REAL WORLD

10. Jerry printed 48 photos. He gave
    4 friends the same number of
    photos. How many photos did each
    friend receive? (pp. P271–P272)

_____

11. Tina divides 17 crayons into 3 equal
    groups. How many crayons will be
    in each group? How many crayons
    will be left over? (pp. P279–P280)

_____

© Houghton Mifflin Harcourt Publishing Company

Fill in the bubble for the correct answer choice.

12. Marita cuts 72 daisies to make bouquets. She makes 6 equal bouquets. How many daisies are in each bouquet? (pp. P273–P274)

    (A) 6   (C) 8

    (B) 7   (D) 12

13. Christine charges $5 an hour to babysit. How much money does she earn in 16 hours? (pp. P277–P278)

    (A) $21   (C) $64

    (B) $50   (D) $80

14. Use the bar graph. Hector divides the carrot seeds evenly in 4 garden plots. How many carrot seeds will be left over? (pp. P279–P280)

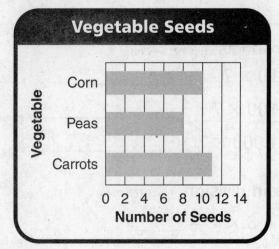

    (A) 5

    (B) 4

    (C) 3

    (D) 2

15. Roberto has 39 model cars. He wants to display an equal number of model cars on each of 3 shelves. How many model cars will he put on each shelf? (pp. P281–P282)

    (A) 2

    (B) 9

    (C) 13

    (D) 39

© Houghton Mifflin Harcourt Publishing Company

Name _____

Model Tenths and Hundredths

Essential Question How can you model and write fractions in tenths and hundredths?

 UNLOCK the Problem

You can use models to represent fractions in tenths and hundredths.

> • What do you need to find to write the fraction?
>
> _____

Example

A

STEP 1

This model has 10 equal parts. Each part is one **tenth**. Shade three parts out of ten equal parts.

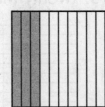

STEP 2

Write the fraction.
Think: Three tenths are shaded.

_____

B

STEP 1

This model has 100 equal parts. Each part is one **hundredth**. Shade eight of one hundred equal parts.

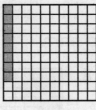

STEP 2

Write the fraction.
Think: Eight hundredths are shaded.

_____

Try This!

Shade the model to show nine of the ten equal parts.

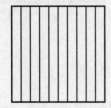

Read: _____

Write: _____

Shade the model to show sixty-five of the hundred equal parts.

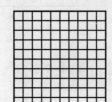

Math Talk Which number in a fraction represents the number of parts being counted, and which represents the number of equal parts in the whole?

Read: _____

Write: _____

© Houghton Mifflin Harcourt Publishing Company

Share and Show

Write the fraction that names the shaded part.

1.    Think: How many equal parts are shaded?

_____

2.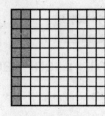

_____

3.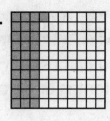

_____

Shade to model the fraction. Then write the fraction in numbers.

4. three tenths

   _____

5. twenty-three hundredths

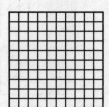   _____

On Your Own . . . .

Write the fraction that names the shaded part.

6.

_____

7.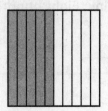

_____

8.

_____

9.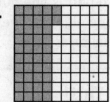

_____

Problem Solving

10. Each player shot a basketball 10 times. Eric made 4 baskets. Write a fraction to represent the part of Eric's shots that were baskets.

_____

11. Nina asked 100 students if they have a pet. Of the students, $\frac{19}{100}$ have a cat. How many students have a cat?

_____

© Houghton Mifflin Harcourt Publishing Company

Name _____

Fractions Greater Than One

Essential Question When might you use a fraction greater than 1 or a mixed number?

**UNLOCK the Problem** REAL WORLD

Troy uses $\frac{1}{4}$ of a box of clay to make one model of a car. How many boxes of clay does he use to make 5 model cars?

- How much clay does Troy use to make each model car?

  _____

- How many model cars does Troy make?

  _____

🔒 **Make a model.**

- Draw squares divided into fourths to show the boxes of clay. Shade $\frac{1}{4}$ for the amount of clay Troy uses for each of the 5 model cars.

- Count the number of shaded parts. There are

  _____ shaded parts.

- Write the fraction.

  ▢ shaded parts

  ▢ parts in the whole

Think: $\frac{4}{4} = 1$

One whole and one fourth are shaded.

Write: $1\frac{1}{4}$

The number $\frac{5}{4}$ is a fraction greater than 1. A fraction greater than 1 can be written as a **mixed number**. A mixed number has a whole number and a fraction.

So, Troy uses $\frac{5}{4}$ or $1\frac{1}{4}$ boxes of clay to make 5 model cars.

Read Math

Read $1\frac{1}{4}$ as *one and one fourth*.

Math Talk Why are $\frac{5}{4}$ and $1\frac{1}{4}$ equal?

© Houghton Mifflin Harcourt Publishing Company

Share and Show

1. Each fraction circle is 1 whole. Write a mixed number for the parts that are shaded.

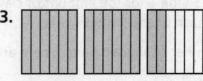

   There are _____ parts shaded.

   There are _____ equal parts in the whole.

   Fraction: ⬚/⬚ shaded parts
   parts in a whole

   There is _____ whole shaded and _____ thirds shaded.

   The mixed number is _____.

Each shape is 1 whole. Write a mixed number for the parts that are shaded.

2.   _____

3. _____

On Your Own

Each shape is 1 whole. Write a mixed number for the parts that are shaded.

4. 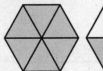 _____

5.  _____

Problem Solving

6. Luis played $\frac{6}{4}$ games of soccer this season. How can you write the number of games Luis played as a mixed number?

_____

7. Marci used $\frac{7}{3}$ packages of juice drinks. How can you write the number of packages of juice drinks Marci used as a mixed number?

_____

© Houghton Mifflin Harcourt Publishing Company

Name _____

Equivalent Fractions

Essential Question How can you use models to find equivalent fractions?

> **UNLOCK the Problem** REAL WORLD

Bart brought an apple pie to the picnic. He cut the pie into 6 equal pieces and 3 pieces were eaten.

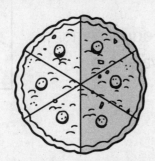

- What fraction names the amount of the pie that

  was eaten? _____

- What fraction names the amount of the pie that

  was left over? _____

Bart divided each of the leftover pieces into 2 equal pieces. Draw a dashed line on each piece to show how Bart divided it.

After you divide each sixth-size piece into 2 equal pieces, there will be 12 pieces in the whole pie. The pieces are called twelfths.

- What fraction names the total number of pieces

  Bart has left? _____

$\dfrac{\square}{6}$ and $\dfrac{\square}{12}$ are equivalent since they both name the same amount of the pie.

> **Math Talk** How do the size of the parts compare in the equivalent fractions? How do the number of parts compare?

© Houghton Mifflin Harcourt Publishing Company

Share and Show

Use models to find the equivalent fraction.

1. $\dfrac{1}{2} = \dfrac{\square}{4}$

This model shows a whole divided into 2 equal parts.
Shade the model to show the fraction $\dfrac{1}{2}$.

This model shows a whole divided into 4 equal parts.
Shade the model to show a fraction equivalent to $\dfrac{1}{2}$.

So, $\dfrac{\square}{2} = \dfrac{\square}{4}$.

On Your Own

Use models to find the equivalent fraction.

2. $\dfrac{1}{2} = \dfrac{\square}{6}$

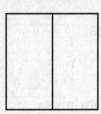
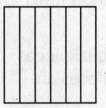

3. $\dfrac{9}{12} = \dfrac{\square}{4}$

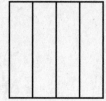

Problem Solving

4. A loaf of bread has 12 slices. Micky ate $\dfrac{1}{4}$ of the loaf. Write the fraction of the loaf Micky ate in twelfths.

5. Sandra used $\dfrac{1}{4}$ of a meter of string to make a bracelet. Write the fraction of a meter of string Sandra used in eighths.

© Houghton Mifflin Harcourt Publishing Company

Name _____

Equivalent Fractions on a Multiplication Table

Essential Question How can you generate equivalent fractions using a multiplication table?

CONNECT You can use a model to show the equivalent fractions $\frac{1}{2}$, $\frac{2}{4}$, and $\frac{3}{6}$.

Think: The same amount is shaded in the models; the second model and third model have more parts shaded.

$$\frac{1}{2} = \frac{2}{4} = \frac{3}{6}$$

🔑 UNLOCK the Problem

You can use a multiplication table for other equivalent fractions for $\frac{1}{2}$.

🔒 Activity What are some equivalent fractions for $\frac{1}{2}$?

Materials ■ multiplication table

- Shade the row for the numerator of the fraction $\frac{1}{2}$. The numerator is 1.

- Shade the row for the denominator of the fraction $\frac{1}{2}$. The denominator is 2.

- Look across the rows for numerator 1 and denominator 2.

- In a multiplication table, how are a product and the product below it related?

_____

_____

×	1	2	3	4	5	6	7	8	9	10
1	1	2	3	4	5	6	7	8	9	10
2	2	4	6	8	10	12	14	16	18	20
3	3	6	9	12	15	18	21	24	27	30

Write the products with the numerator 1 as a factor. Then write the products with the denominator 2 as a factor. The first three are done for you.

numerator ⟶
denominator ⟶
$$\frac{1}{2} = \frac{2}{4} = \frac{3}{6} = \frac{}{8} = \frac{6}{}$$

Math Talk Why is the arrangement of factors and products in a multiplication table helpful in finding equivalent fractions?

- What do you notice about the products from the column for 1 to the column for 2?

  The numerator and denominator both increase by a factor of _____.

- What do you notice about the products from the column for 1 to the column for 3?

  The numerator and denominator both increase by a factor of _____.

- What do you notice about the products from the column for 1 to the column for 4?

  The numerator and denominator both increase by a factor of _____.

Math Idea

To find an equivalent fraction, you can multiply both the numerator and denominator by the same number.

So, $\frac{2}{4}$, $\frac{3}{6}$, $\frac{4}{8}$, and $\frac{6}{12}$ are some equivalent fractions for $\frac{1}{2}$.

© Houghton Mifflin Harcourt Publishing Company

Share and Show

Use a multiplication table to find equivalent fractions.

1. Write 3 equivalent fractions for $\frac{1}{3}$.

×	1	2	3	4	5	6	7	8	9	10
1	1	2	3	4	5	6	7	8	9	10
2	2	4	6	8	10	12	14	16	18	20
3	3	6	9	12	15	18	21	24	27	30

   • Shade the row for the numerator of the
   fraction $\frac{1}{3}$. The numerator is _____.

   • Shade the row for the denominator of the
   fraction $\frac{1}{3}$. The denominator is _____.

   • Look across the rows for numerator 1 and denominator 3.
   Write the products with the numerator 1 as a factor. Then write
   the products with the denominator 3 as a factor.

   numerator ⟶ $\frac{1}{3} = \frac{}{6} = \frac{}{} = \frac{}{}$.
   denominator ⟶

   So, $\frac{1}{3} = \frac{}{} = \frac{}{} = \frac{}{}$.

List 3 equivalent fractions.

2. $\frac{1}{6}$

   _____

3. $\frac{1}{4}$

   _____

On Your Own

Use a multiplication table to find three equivalent fractions.

4. $\frac{2}{5}$

   _____

5. $\frac{3}{10}$

   _____

Problem Solving REAL WORLD

6. On Jan's soccer team, $\frac{1}{5}$ of the
   players are on the field. What are
   three equivalent fractions that name
   the part of the team on the field?

7. Chen used $\frac{3}{4}$ of a carton of milk.
   What are three equivalent fractions
   that name the part of the carton of
   milk that Chen used?

_____

_____

© Houghton Mifflin Harcourt Publishing Company

Name _____

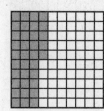 **Checkpoint**

Concepts and Skills

Write the fraction that names the shaded part. (pp. P285–P286)

1.

_____

2.

_____

Each shape is 1 whole. Write a mixed number for the parts that are shaded. (pp. P287–P288)

3.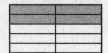

_____

Use models to find the equivalent fraction. (pp. P289–P290)

4. $\dfrac{1}{4} = \dfrac{}{12}$ ⊕ ✳

5. $\dfrac{5}{6} = \dfrac{}{12}$ ⊗ ✳

Use a multiplication table to find three equivalent fractions. (pp. P291–P292)

6. $\dfrac{3}{4}$

_____

7. $\dfrac{4}{10}$

_____

Problem Solving REAL WORLD

8. Three friends shared 4 pies equally. Each person got $\frac{4}{3}$ pies. How can you write how much pie each person got as a mixed number?

_____

9. Bill bought a large submarine sandwich and cut it into 8 equal pieces. He ate $\frac{1}{4}$ of the sandwich. How can you write how much of the sandwich Bill ate as eighths?

_____

© Houghton Mifflin Harcourt Publishing Company

Fill in the bubble for the correct answer choice.

10. Each player hit a baseball 10 times. Linda batted 8 balls to the outfield. Write a fraction to show what part of 10 hits Linda batted to the outfield. (pp. P285–P286)

   Ⓐ $\frac{18}{18}$

   Ⓑ $\frac{10}{8}$

   Ⓒ $\frac{9}{10}$

   Ⓓ $\frac{8}{10}$

11. Vilma used $\frac{8}{3}$ packages of graham crackers to make piecrusts. How can you write the packages of crackers Vilma used as a mixed number? (pp. P287–P288)

   Ⓐ $2\frac{1}{8}$ Ⓒ $2\frac{2}{3}$

   Ⓑ $2\frac{1}{3}$ Ⓓ $3\frac{1}{3}$

12. Sam used $\frac{10}{12}$ of a meter of ribbon to decorate a picture frame. What fraction of a meter of ribbon, in sixths, did Sam use? (pp. P289–P290)

   Ⓐ $\frac{2}{12}$

   Ⓑ $\frac{5}{6}$

   Ⓒ $\frac{6}{12}$

   Ⓓ $\frac{12}{10}$

13. Leona used $\frac{3}{8}$ of a bottle of juice. Which is an equivalent fraction that names the part of the bottle of juice that Leona used? (pp. P291–P292)

   Ⓐ $\frac{6}{16}$ Ⓒ $\frac{3}{4}$

   Ⓑ $\frac{5}{8}$ Ⓓ $\frac{8}{3}$

© Houghton Mifflin Harcourt Publishing Company

Name _____

Same Size, Same Shape

Essential Question How can you identify shapes that have the same size and are shaped the same?

🔑 UNLOCK the Problem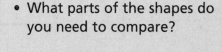

You can tell if two shapes have the same size and are shaped the same by comparing the matching parts of the shapes.

- What parts of the shapes do you need to compare?

_____

🔑 Activity Compare size and shape.

Materials ■ grid paper ■ scissors ■ ruler

STEP 1 Trace Shape A on grid paper. Cut out Shape A.

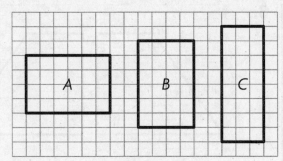

STEP 2 Move Shape A in any way to compare it to Shape B.

- Do the shapes match exactly? _____

Shape A and Shape B _____ the same

size and _____ shaped the same.

STEP 3 Move Shape A in any way to compare it to Shape C.

- Do the shapes match exactly? _____

Shape A and Shape C _____ shaped the same.

Try This!

Since all the angles in Shapes A and B are the same, you can compare shapes by their matching sides.

The length of the shorter side of Shape A is _____ units.

The length of the shorter side of Shape B is _____ units.

The length of the longer side of Shape A is _____ units.

The length of the longer side of Shape B is _____ units.

So, Shape A and Shape B have the _____ size and are

shaped the _____.

Math Talk **Explain** how the size and shape of Shape A compares to the size and shape of Shape C.

© Houghton Mifflin Harcourt Publishing Company

Share and Show

1. Which shape appears to have the same size and the same shape as Shape *A*?

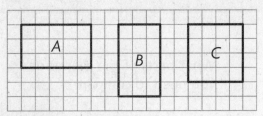

Think: If I trace Shape *A* and move it, which shape might it match exactly?

_____

On Your Own ....

Look at the first shape. Tell if it appears to have the same size and shape as the second shape. Write *yes* or *no*.

2.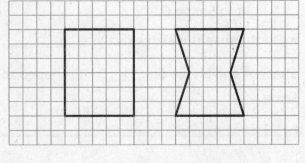

_____

3.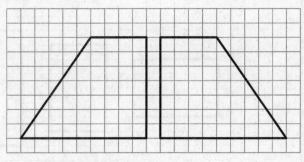

_____

4.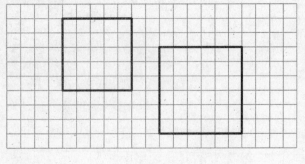

_____

5.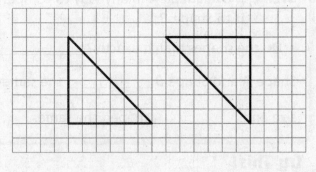

_____

Problem Solving

6. Kyra says that these shapes have the same size and same shape. Is she correct? **Explain**.

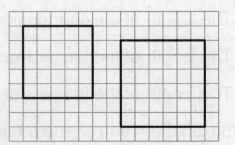

_____

_____

© Houghton Mifflin Harcourt Publishing Company

Name _____

Change Customary Units of Length

Essential Question How can you change feet to inches?

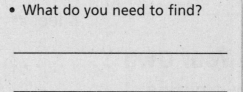

🔑 UNLOCK the Problem REAL WORLD

You can use different units to name the same length.

Erin has a shelf that is 2 feet long. How many inches long is Erin's shelf?

• What do you need to find?

_____

_____

🔒 One Way Draw a picture.

2 feet

1 foot	1 foot

Remember
1 foot = 12 inches

Draw one box to show each foot. Below each foot, draw 12 small boxes to show the number of inches in 1 foot. Count the total number of small boxes.

There are 24 small boxes in all. 2 feet = _____ inches.

So, Erin's shelf is _____ inches long.

🔒 Another Way Use a number line.

Erin has a table that is 3 feet long. How many inches long is her table? Draw a number line and label it in feet.

_____ in.   +   _____ in.   +   _____ in.

0 ft   1 ft   2 ft   3 ft   4 ft

Draw a 12-inch jump for each foot. Add the lengths of the jumps to find the total number of inches.

Math Talk Why do you count by 12s when you rename feet as inches?

3 feet = _____ inches.

So, Erin's table is _____ inches long.

© Houghton Mifflin Harcourt Publishing Company

Share and Show

1. Use the number line. Rename 4 feet using inches.

0 ft   1 ft   2 ft   3 ft   4 ft   5 ft

4 feet = _____ inches

On Your Own ·

Draw a picture.

2. Rename 7 feet using inches.

7 feet = _____ inches

3. Rename 6 feet using inches.

6 feet = _____ inches

4. Use the number line. Rename 8 feet using inches.

8 feet = _____ inches

Problem Solving

5. Ella has a rope that is 10 feet long. How many inches long is the rope?

_____

6. Jose is 5 feet tall. How many inches tall is he?

_____

P298

© Houghton Mifflin Harcourt Publishing Company

Change Metric Units of Length

Essential Question How can you change meters to centimeters?

CONNECT You have learned to change feet to inches. In this lesson, you will change meters to centimeters.

🔓 UNLOCK the Problem REAL WORLD

Gina needs a piece of wood that is 4 meters long to make a bench. How many centimeters of wood does Gina need?

- What do you need to do to answer the question?
  _____

Complete the table to show how the units are related.

STEP 1 Look for a pattern to complete the table. Describe the relationship.

Remember
1 meter = 100 centimeters

Meters	1	2	3	4	5
Centimeters	100	200	300	**400**	

To find the number of centimeters, add _____ centimeters for each meter.

STEP 2 Use the relationship to find the number of centimeters in 4 meters.

4 meters = _____ centimeters

So, Gina needs _____ centimeters of wood to make a bench.

Examples

A. Change 6 meters to centimeters.

Add 100 to _____ centimeters.

So, 6 meters = _____ centimeters.

B. Change 8 meters to centimeters.

Multiply 100 centimeters by _____.

So, 8 meters = _____ centimeters.

Math Talk What do you need to know in order to change from one unit of length to another?

© Houghton Mifflin Harcourt Publishing Company

Share and Show

1. How can you change 3 meters to centimeters?
Complete the table to show how the units are related.

Meters	1	2	3	4
Centimeters	100	200		400

To find the number of centimeters,

add _____ centimeters for each meter.

So, 3 meters = _____ centimeters.

Find the unknown number.

2. 2 meters = _____ centimeters

3. 5 meters = _____ centimeters

On Your Own ·

Complete the table.

4.

Meters	3	4	5	6	7	8	9	10
Centimeters	300	400	500				900	

Find the unknown number.

5. 8 meters = _____ centimeters

6. 3 meters = _____ centimeters

Problem Solving

7. Jorge needs 7 meters of wire for a garden fence. The wire is sold in centimeters. How many centimeters of wire does Jorge need?

8. Wanda needs 9 meters of fabric to make curtains. She has 1,000 centimeters of fabric. Does Wanda have enough fabric to make the curtains? **Explain**.

_____

_____

© Houghton Mifflin Harcourt Publishing Company

Name _____

Estimate and Measure Liquid Volume

Essential Question How are cups, pints, quarts, and gallons related?

🔑 UNLOCK the Problem

You can use customary units to measure the amount of liquid a container will hold. Some customary units are **cup (c)**, **pint (pt)**, **quart (qt)**, and **gallon (gal)**.

1 cup (c)	1 pint (pt)	1 quart (qt)	1 gallon (gal)

🔑 Activity Show how cups, pints, quarts, and gallons are related.

Materials ■ cup, pint, quart, gallon containers ■ water

STEP 1 Estimate the number of cups it will take to fill the pint container. Record your estimate in the table.

STEP 2 Fill a cup and pour it into the pint container. Repeat until the pint container is full. Record the number of cups it took to fill the pint container.

STEP 3 Repeat Steps 1 and 2 for the quart and gallon containers.

Number of Cups			
	Number of Cups in a Pint	Number of Cups in a Quart	Number of Cups in a Gallon
Estimate			
Liquid Volume			

Math Talk Which unit would you use to measure the amount of water needed to fill an aquarium? **Explain** your choice.

© Houghton Mifflin Harcourt Publishing Company

Share and Show

Choose the unit you would use to measure the amount of liquid the container will hold. Write *cup, pint, quart,* or *gallon.*

1.    **Think:** A cup is small.

   cup
   _____

2.

   bucket

   _____

3.

   bathtub

   _____

4.

   glass

   _____

On Your Own

Choose the unit you would use to measure the amount of liquid the container will hold. Choose the better unit of measure.

5. a dog's water bowl: 2 cups or 2 gallons

6. a juice box: 1 cup or 1 quart

Problem Solving

7. Lila made 3 quarts of lemonade. How many cups of lemonade did she make?

   _____

8. Richard made 2 gallons of fruit punch for a party. How many 1-cup servings can he make?

   _____

© Houghton Mifflin Harcourt Publishing Company

Name _____

Estimate and Measure Weight

Essential Question How are ounces and pounds related?

🔓 UNLOCK the Problem   REAL WORLD

Weight is the measure of how heavy an object is. Customary units of weight include **ounce (oz)** and **pound (lb)**.

Customary Units of Weight
1 pound = 16 ounces

1 slice of bread weighs about 1 ounce.

1 loaf of bread weighs about 1 pound.

🔑 Activity Show how ounces and pounds are related.

Materials ■ spring scale ■ classroom objects

STEP 1 Estimate the weight of the object shown in the table. Record your estimate.

_____

STEP 2 Use a scale to measure the weight of the object to the nearest ounce or pound. Record the weight.

_____

STEP 3 Repeat Steps 1 and 2 for each object.

Remember

Include the unit when you record each estimate and measurement in your table.

Weight of Objects		
Object	Estimate	Weight
apple		
book		
pencil box		
tape dispenser		

Math Talk How do your estimates compare to the actual weights?

© Houghton Mifflin Harcourt Publishing Company

Share and Show

1. Which unit would you use to measure the weight of a grape? Write *ounce* or *pound*.

   Think: A grape is a small, light object.

   <u>ounce</u>

Choose the unit you would use to measure the weight. Write *ounce* or *pound*.

2.

_____

3.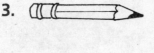

_____

4

_____

On Your Own

Choose the unit you would use to measure the weight. Write *ounce* or *pound*.

5.

_____

6.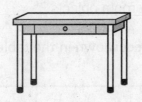

_____

7

_____

Problem Solving REAL WORLD

8. Duane bought some oregano to use in a batch of pasta sauce. Which is a more likely weight for the oregano, 1 ounce or 1 pound?

_____

9. Erin bought a bag of flour to use for baking dinner rolls. Did she buy 5 ounces of flour or 5 pounds of flour?

_____

© Houghton Mifflin Harcourt Publishing Company

Name _____

✓ Checkpoint

Concepts and Skills

Look at the first shape. Tell if it appears to have the same size
and shape as the second shape. Write *yes* or *no*. (pp. P295–P296)

1.
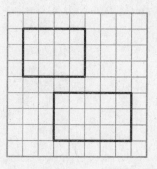

_____

2.
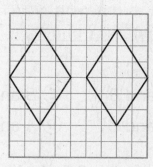

_____

3. Use the number line. Rename 5 feet using inches. (pp. P297–P298)

←——————————————————————————————→

5 feet = _____ inches

Find the unknown number. (pp. P299–P300)

4. 6 meters = _____ centimeters

5. 8 meters = _____ centimeters

Choose the unit you would use to measure the amount of liquid
the container will hold. Choose the better unit of measure. (pp. P301–P302)

6. a pitcher of iced tea: 1 cup or 1 gallon

Problem Solving REAL WORLD

7. A tea pot holds 4 quarts of tea.
   How many 1-cup servings of tea
   does it hold? (pp. P301–P302)

8. Evan bought a large bag of dry
   dog food for his dog. Did Evan
   buy 6 ounces or 6 pounds of dog
   food? (pp. P303–P304)

_____   _____

© Houghton Mifflin Harcourt Publishing Company

Fill in the bubble for the correct answer choice.

9. Which shapes appear to have the same size and shape? (pp. P295–P296)

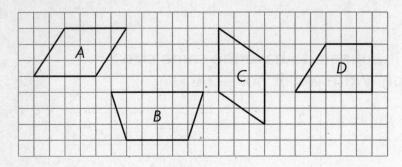

   Ⓐ *A* and *B*   Ⓒ *B* and *D*

   Ⓑ *B* and *C*   Ⓓ *A* and *C*

10. Trey's desk is 3 feet wide. How many inches wide is the desk? (pp. P297–P298)

   Ⓐ   3 inches   Ⓒ 36 inches

   Ⓑ 24 inches   Ⓓ 48 inches

11. Juana needs 2 meters of yarn for a friendship bracelet. How many centimeters of yarn does she need? (pp. P299–P300)

   Ⓐ 2,000 centimeters   Ⓒ 20 centimeters

   Ⓑ   200 centimeters   Ⓓ   2 centimeters

12. Lana made 3 quarts of soup. How many pints of soup did she make? (pp. P301–P302)

   Ⓐ   6 pints   Ⓒ 18 pints

   Ⓑ 12 pints   Ⓓ 24 pints

13. Which object weighs about 1 ounce? (pp. P303–P304)

   Ⓐ a loaf of bread   Ⓒ a strawberry

   Ⓑ a watermelon   Ⓓ a chair

© Houghton Mifflin Harcourt Publishing Company